RENEWALS 458-4574

GEOMORPHOLOGY AND ENVIRONMENTAL IMPACT ASSESSMENT

Geomorphology and Environmental Impact Assessment

Edited by
MAURO MARCHETTI
Dipartimento di Scienze della Terra, Università degli Studi di Modena e Reggio Emilia, Italy

VICTORIA RIVAS
Departamento de Geografía, Urbanismo y Ordenación del Territorio, Universidad de Cantabria, Spain

A.A. BALKEMA PUBLISHERS   LISSE / ABINGDON / EXTON (PA) / TOKYO

Cover illustration: October 2000, Flood at the R. Mincio - R. Po confluence (Northern Italy)
Cover design: Studio Jan de Boer, Amsterdam, The Netherlands
Printed by : Grafisch Produktiebedrijf Gorter, Steenwijk, The Netherlands

Published by
A.A. Balkema Publishers, a member of Swets & Zeitlinger Publishers
www.balkema.nl and www.szp.swets.nl

© 2001 Swets & Zeitlinger B.V.

All rights reserved. No part of this publication may be reproduced, stored in a retrieval system, or transmitted in any form or by any means: electronic, mechanical, by photocopying, recording or otherwise, without the prior written permission of the publishers.

ISBN 90 5809 344 1

Table of contents

Foreword   vii

Geomorphology and Environmental Impact Assessment   1
A. Cendrero, M. Marchetti, M. Panizza, and V. Rivas

Geomorphological impact assessment in the River Mincio plain (Province of Mantova,
Northern Italy)   7
F. Baraldi, D. Castaldini, and M. Marchetti

Prediction models for landslide hazard zonation using a fuzzy set approach   31
C.F. Chung and A.G. Fabbri

Impact of land-recycling projects: Evaluation and visualisation of risk assessments   49
K. Heinrich

Geomorphology and Environmental Impact Assessment: A case study in Moena
(Dolomites - Italy)   71
M. Marchetti and M. Panizza

Basics of the natural river engineering   83
D. Komatina

Impacts of natural salt transport in a coastal Mediterranean environment   113
J. Bach, J. Trilla, and R. Linares

Environmental reclamation of dismissed quarries in protected areas:
A case study near Rome (Italy)   123
G. Di Filippo, M. Pecci, F. Silvestri, and F. Biondi

Landscape analysis based on environmental units and visual areas. The use of
geomorphological units as a basic framework. La Vall de Gallinera, Alicante (Spain)   133
I. de Villota, J.L. Goy, C. Zazo, I. Barrera, and J. Pedraza

The use of models for the assessment of the impact of land use on runoff production
in the Ouveze catchment (France)   155
Th.W.J. van Asch and S.J.E. van Dijck

Floods in the city of Rio Branco, Brazil: A case study on the impacts of human
activities on flood dynamics and effects   163
E.M. Latrubesse, S. Aquino da Silva, M. de J. Moraes

Geomorphological impacts and feasibility of laying-out ski-trails in a glacial cirque
in the Vosges (France) using GIS and multi-criteria analyses   177
S.J.E. van Dijck

A methodology for assessing landscape quality for environmental impact assessment
and land use planning: Application to a Mediterranean environment   191
L. Recatalá and J. Sánchez

GIS applications for Environmental Impact Assessment: An analysis of the effects of
a motor way project in a alpine valley   207
A. Patrono, F. de Francesch, M. Marchetti, and A. Moltrer

Foreword

Most of the contributions of this volume were presented in Granada at a Special Symposium entitled: "Geomorphology and Environmental Impact Assessment" within the framework of the 6th Spanish Congress and International Conference on Environmental Geology and Land-use Planning.

The papers address some of the issues concerning the relationships between Geomorphology and Environmental Impact Assessment as mentioned in the first paper of this volume by Cendrero et al., although by no means all of them. They provide a good overview of a variety of methods and approaches, often of an interdisciplinary character and with an emphasis on the integration of geomorphological analyses into the general process of environmental evaluation.

The volume has been edited by two young researchers, Mauro Marchetti and Victoria Rivas, who have been collaborating in several research projects on geomorphological implication in the Environmental Impact and have acquired a considerable experience in this study sector.

A. Cendrero and M. Panizza

Geomorphology and Environmental Impact Assessment

A. Cendrero
DCITIMAC, Universidad de Cantabria, Santander, Spain

M. Marchetti & M. Panizza
Dipartimento di Scienze della Terra, Università degli Studi di Modena, Italy

V. Rivas
Departamento de Geografía, Urbanismo y Ordenación del Territorio, Universidad de Cantabria, Santander, Spain

Environmental Impact Assessment (EIA) is a very topical subject both for the implementation of considerably important engineering works and for territorial planning. As a consequence, nowadays many researchers are involved in this important scientific branch which is introduced more and more often as an indispensable procedure in the laws and regulations of several countries. The scientific contributions collected in this volume show how widespread this topic is geographically.

When a project is undertaken, there may occur several effects with environmental consequences. These effects may be direct or indirect, permanent or temporary, single or cumulative, short- or long-term. Not only are effects induced in the operative phase but also in the decommissioning phase (running down and dismantling).

Seldom are the effects on the environment sufficiently considered in the general planning. The operations and steps of Environmental Impact Assessment (EIA) studies are summarised in Figure 1 (Panizza 1995). When operations are functioning it is recommended that the degree of success of the predictions and impacts be evaluated as a guide to maintenance routines and future design. In the decommissioning phase it will be necessary to establish a new EIA, in order to predict the effects of the abandonment or dismantling of the plant (i.e. widespreading of pollutants, uncontrolled mine drainage etc.).

Although in some European countries EIA studies are compulsory, the geomorphological component is generally not assessed explicitly (Barani et al. 1995, Rivas et al. 1995, van Asch & van Dijck 1995). Generic considerations and geomorphological characteristics are however usually included in other environmental components (Watern 1990).

There are various reasons for the limited presence of geomorphology in EIA studies. First of all, geomorphology is not considered important by non-specialists. Many unmistakable and dramatic examples could be quoted to support this idea. The intense urban development of debris and alluvial cones in the Alpine region, or along the Blue River banks in China and in very many other examples all over the world, prove that man has a short memory with respect to processes which occur periodically, with varying return times, and cause risk situations for man and his activities.

Secondarily, non-specialists, but often also decision makers and the general public, tend to consider the landscape as a permanent set-up in a static equilibrium rather than the result of complex processes in a dynamic equilibrium which can also imply peaks of intense activity.

Furthermore up to not long ago earth science specialists showed superficial interest in the general environment. Only in the past two decades have new initiatives been carried out to improve our understanding of all the phenomena interacting in a natural environment and which, therefore, also influence geomorphological features. This has coincided with the growth of awareness among the general public that our geomorphological heritage must be preserved and that hazardous geomorphological processes which put human life and activities at risk must be assessed more accurately (Cooke & Doornkamp 1990, Panizza 1996).

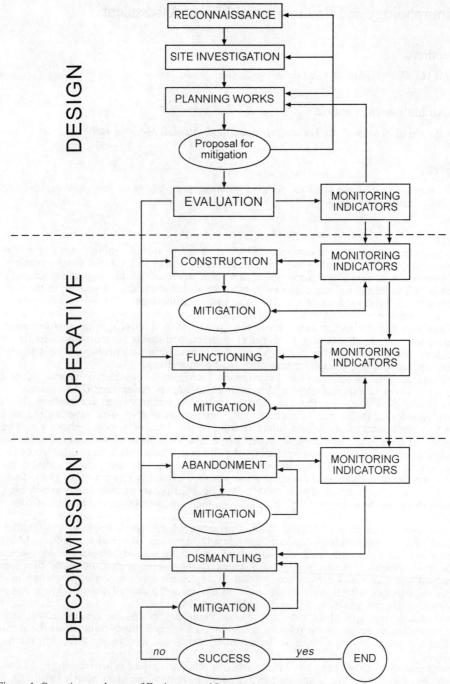

Figure 1. Operations and steps of Environmental Impact Assessment studies.

In a project financed by the European Union, the fields of research, the definitions and procedures to be developed for the introduction of the geomorphological component in EIA studies have been discussed (Marchetti et al. 1995, Panizza et al. 1995). In particular, three different geomorphological components have been identified and treated separately (Fig. 2).

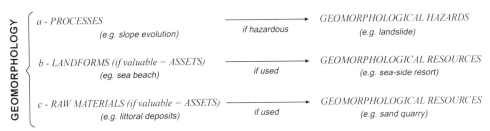

Figure 2. Main groups of geomorphological components.

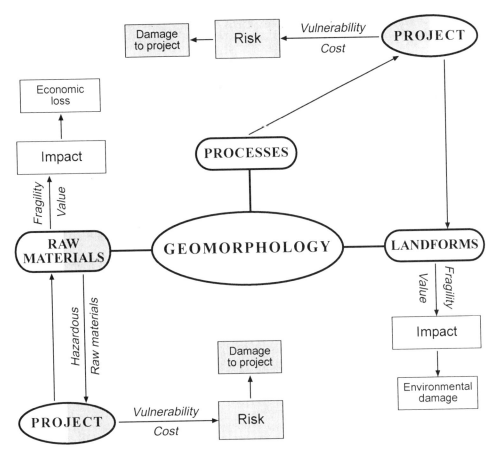

Figure 3. Conceptual basis for the relationships between geomorphological components, raw materials, processes and a project.

Figure 3 summarises the conceptual basis of the simplified relationships between these geomorphological components and a project.

These concepts have been presented in more detail by Cavallin et al. (1994) and Panizza (1995, 1996). It may be noticed that when processes are hazardous, they become geomorphological hazards and if they interfere with a project having its own cost and vulnerability, they may cause damage to the project itself. In this case natural geomorphological processes play an active role whereas the project plays a passive one.

When particular landforms are valuable, they are considered geomorphological assets, since they are characterized by a specific fragility and value, and these may be affected by a project. Landforms may be regarded as valuable assets not only on the basis of scenic, social, economic and cultural values, but also on the basis of other criteria related to the more general scientific concept of asset.

The methods for identifying and evaluating geomorphological assets on a scientific basis may be subdivided into two phases: a) geomorphological survey and mapping; b) selection from geomorphological maps of those landforms that may be considered as assets. The implementation of a project can have a direct impact on geomorphological assets and cause environmental damage to them. In this case the natural component of the physical environment plays a passive role whereas the project plays a direct one.

The same remarks can be made for particular raw materials: when these are valuable, they are considered geomorphological assets and the interference associated with a project may produce a direct impact. In this situation there are some materials that may also be considered as hazards (i.e. salty soils, radioactive minerals, metastable sands, and so on). They may therefore create a risk for the project and produce damage to it.

Another reason which does not favour the introduction of the geomorphological component in EIA studies is the limited development of specific tools usable to quantify the assessment of the interactions occurring between landscape, geomorphological processes and human activities. There are basically two reasons for this. The first is that modern geomorphology is a very young branch of earth sciences in which a lot of instruments, already utilized in other disciplines, are only being introduced. The second and more important reason is that geomorphology is not a systematic discipline and the physical laws applicable in general cannot be used in a specific case but must be modified to respond to local causes, specific parameters and human activities which do not always correspond in different countries, and so on.

In some cases it is possible to use modelling or parametric approaches in order to predict the consequences of certain interventions (Amarù & van Asch 1996, Amarù et al. 1996, Rivas et al. 1996). However, in most instances, predictions may only be formulated by means of general, qualitative terms. This is in part due to the limited knowledge of the role of different factors in the development of geomorphological processes and in part to the fact that in most EIA studies the necessary baseline data are not available or cannot be obtained owing to limited time and/or resources.

The need to understand natural processes, on the one hand, and to identify the effects of potential modifications introduced by human activity, on the other, are the main goals of any EIA study. Several steps are required to carry out an environmental impact study, including: a) formulation of alternatives; b) impact identification; c) description of the existing environment; d) prediction and assessment of the impacts; e) selection of the proposed action from a set of previously evaluated alternatives.

In recent years, studies focusing on the role of spatial information systems in regional and urban planning have been carried out. These investigations required either the design and use of tailor-made information systems or of GISs which were particularly well suited for spatial modelling.

The characterisation of geomorphological impacts, resources and assets for environmental impact assessment should therefore be reconsidered. The approach is to facilitate computational procedures by means of geographic information systems. Spatial representation and spatial normalization are fundamental steps in the construction of a balanced framework for assessment, since their relevance extends far beyond the morphological area of environmental applications.

REFERENCES

Amarù, M. & van Asch, T.W.J. 1996. Modelling environmental impacts of land-use changes on overland flow and channel flow processes. In M. Panizza, A.G. Fabbri, M. Marchetti & A. Patrono (eds), Special EU Project Issue: Geomorphology and Environmental Impact Assessment. *ITC Journal*, 1995(4): 327-330.

Amarù, M., van Dijck, S.J.E. & van Asch, T.W.J. 1996. Methods for environmental impact assessment related to geomorphic processes. In M. Panizza, A.G. Fabbri, M. Marchetti & A. Patrono (eds), Special EU Project Issue: Geomorphology and Environmental Impact Assessment. *ITC Journal*, 1995(4): 325-327.

Barani, D., Ferrari G., Pergreffi, G. & Vezzani, A. 1995. The role of earth sciences in the environmental impact assessment in Italy. In M. Marchetti, M. Panizza, M. Soldati & D. Barani (eds), Geomorphology and Environmental Impact Assessment. Proc. 1srt and 2nd workshops of a "Human Capital and Mobility" project. *Quad. Geodinamica Alpina e Quaternaria*, 3: 77-81.

Cavallin, A., Marchetti, M., Panizza, M. & Soldati, M. 1994. Te role of geomorphology in environmental impact assessment. *Geomorphology*, 9: 143-153.

Cooke, V. & Doornkamp, J.C. 1990. *Geomorphology in Environmental Management*. Oxford: Clarendon.

Marchetti, M., Panizza, M., Soldati, M. & Barani, D. (eds) 1995. Geomorphology and Environmental Impact Assessment. Proc. 1st and 2nd workshops of a "Human Capital and Mobility" project. *Quad. Geodinamica Alpina e Quaternaria,* 3: 1-197.

Panizza, M 1995. Introduction to a research methodology for Geomorphology and Environmental Impact Assessment. In M. Marchetti, M. Panizza, M. Soldati & D. Barani (eds), Geomorphology and Environmental Impact Assessment. Proc. 1srt and 2nd workshops of a "Human Capital and Mobility" project. *Quad. Geodinamica Alpina e Quaternaria*, 3: 13-26.

Panizza, M. 1996. *Environmental Geomorphology*. Amsterdam: Elsevier.

Panizza, M., Fabbri, A.G., Marchetti, M. & Patrono, A. (eds) 1996. Special EU Project Issue: Geomorphology and Environmental Impact Assessment. *ITC Journ.*, 1995(4): 305-371.

Rivas, V., Rix, K., Francés, E., Brunsden, D., Cendrero, A. & Collison, A. 1995. Geomorphology and EIA in Spain and Great Britain; a brief review of legislation and practice. In M. Marchetti, M. Panizza, M. Soldati & D. Barani (eds), Geomorphology and Environmental Impact Assessment. Proc. 1srt and 2nd workshops of a "Human Capital and Mobility" project. *Quad. Geodinamica Alpina e Quaternaria*, 3: 83-97.

Rivas, V., Cendrero, A. & Francés, E. 1996. A parametric approach for modelling human impact on coastal cliff erosion. In M. Panizza, A.G. Fabbri, M. Marchetti & A. Patrono (eds), Special EU Project Issue: Geomorphology and Environmental Impact Assessment. *ITC Journ.*, 1995(4): 337-338.

van Asch, T.W.J. & van Dijck, S.J.E. 1995. The role of geomorphology in environmental impact assessment in the Netherlands. In M. Marchetti, M. Panizza, M. Soldati & D. Barani (eds), Geomorphology and Environmental Impact Assessment. Proc. 1srt and 2nd workshops of a "Human Capital and Mobility" project. *Quad. Geodinamica Alpina e Quaternaria*, 3: 63-70.

Watern, P. 1990. The EIA directive of the European Community. In P. Watern (ed), *Environmental Impact Assessment: Theory and Practice*: 192-209. London: Routledge.

Geomorphological impact assessment in the River Mincio plain (Province of Mantova, Northern Italy)

F. Baraldi
U.O. 4.8 Gruppo Nazionale Difesa dalle Catastrofi Idrogeologiche - C.N.R., Italy

D. Castaldini & M. Marchetti
Dipartimento di Scienze della Terra, Università degli Studi di Modena, Italy

ABSTRACT: This study deals with changes in the landscape caused by human activity in the past 40 years in the Province of Mantova in a sector of the River Mincio plain located between the morainic hills of Lake Garda to the north and the Mantova lakes to the south. The study includes the analysis of topographical maps and of aerial photographs taken in different years, and the compilation of geomorphological maps referring to the landscape features in 1997 and in 1955. The plain sector considered has been subjected to intense quarrying activities since the beginning of this century. The open quarries are classified as quarries exploited above the water table and trench quarries exploited below the water table. When no longer in use, the first-type quarries are reclaimed for farming after laying a pedogenised level of organic soil on their floor; the second-type quarries are abandoned or equipped for recreational fishing or in some cases used as occasional dumping sites. All these quarrying activities have caused significant landscape changes. Another aspect of human activities during the past decades concerns the construction of important artificial canals which, besides modifying the natural flow of both surface and sub-surface waters, have altered the natural morphological features of the areas affected. Urban development has also been responsible for important modifications in the landscape, especially in the vicinity of the main built-up areas. In this research a simple methodology for the assessment of the scientific quality of landforms was applied. The impact on the landscape is defined as the reduction in scientific quality due to the assessment of the deterioration produced by human activity. The study has shown that among the main human activities the most serious damage to the landscape has been caused by quarrying. The greatest impact, with over 50% loss of quality of the geomorphological assets, has occurred between Goito and Rivalta.

1 INTRODUCTION

This study deals with landscape modifications induced by human activity in the past 40 years in the Mantova Province in a sector of the River Mincio plain located between the Pleistocene morainic hills of Lake Garda to the north and the Mantova lakes to the south (Fig. 1). Studies performed in past years in the area of the central Po Plain have shown that the River Mincio Plain north of Mantova represents an area in which, more than elsewhere, human activity has played a fundamental role in modelling the physical landscape (Baraldi et al. 1976, 1980a, Baraldi & Zavatti 1990, Dal Ri 1991, Baraldi & Zavatti 1994, Castaldini 1994, Guzzetti et al. 1996, Castiglioni 1997a, b). Nevertheless, only in a few papers (Baraldi et al. 1980c, 1987) the problem of landscape degradation due to human activities has been considered. Therefore the aim of this study was to carried out a geomorphological impact assessment in the River Mincio Plain, north of Mantova. Since human impacts in the Mantova Province have assumed a great importance after the second world war, this study is based essentially on comparison between the morphological situation of the mid-1950s, as evidenced by maps and aerial photographs of that time, and the present-day geomorphological features determined on the basis of recent maps and aerial photographs as well as detailed field surveys.

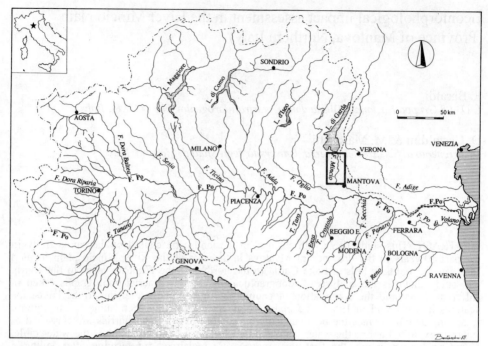

Figure 1. The Po River basin and the study area.

2 GEOGRAPHICAL AND GEOLOGICAL OUTLINE

The sector of the plain under study has an area of about 260 km² and belongs to the Po Plain which is the most extensive plain in Italy (approximately 46,000 km²). The River Mincio, which is the main stream of the study area, is the collector between Lake Garda and the Po River, into which it empties after a course of about 75 Km (Fig. 1).

From a climatic point of view, the Po Plain is situated in a temperate zone (Type C, Koppen's classification). The study area represents one of the least rainy sectors within the Po plain with average annual rainfall from about 750 mm (in the foothill area) to 650 mm (in the Mantova sector), with seasonal peaks concentrated in the autumn and spring, and minimum amounts in the summer and winter (Baraldi & Zavatti 1994). As regards temperature, the data show a mean annual temperatures of 13-14°C (Baraldi & Zavatti 1994).

From a geomorphological point of view, immediately north of the study area is located the morainic amphitheatre of Lake Garda corresponding to a hilly landscape (Fig. 2); here glacial deposits of the Upper Pleistocene crop out (Venzo 1965). Morainic remnants belonging to the Middle Late Pleistocene crop out to the west, near the River Chiese, with small relief emerging at a few metres height with respect to the surrounding plain (Cremaschi 1987). The plain in front of the morainic hills corresponds to the outwash plain (sandur) of the Upper Pleistocene deposited by the spills of the Garda glacier (Cremaschi 1987). The outwash plain, which is also called "Main level of the plain" (Petrucci & Tagliavini 1969, Guzzetti et al. 1996), is terraced by streams of Alpine origin such as the Chiese, Oglio and Mincio rivers. South of the outwash plain lies the alluvial plain of the Po, of Holocene age.

The study area comprises the outwash plain and the wide triangular depression (about 20 km from north to south and 8 km from west to east) cut by the River Mincio, north of the city of Mantova. The morphology of this depression is characterised by several scarps of varying height, mainly developing in a N-S direction and forming various orders of terraces of Holocene age. The natural morphology of the study area has been extensively altered by human intervention, especially quarrying.

At the southern edge of the study area are the Mantova Lakes, into which and out of which the River Mincio flows. These have an area of over 6 km² and a mean depth of 3.50

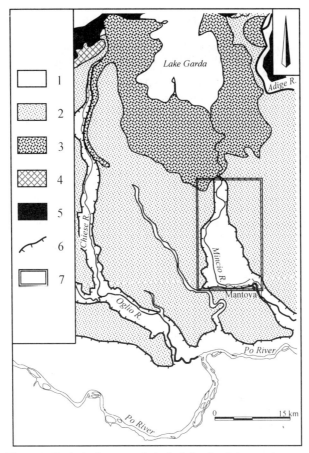

Figure 2. Geological-geomorphological sketch of the area between Lake Garda and the Po River. 1) Alluvial plain (Holocene), 2) Outwash plain (Late Pleistocene), 3) Morainic hills (Upper-Middle Pleistocene), 4) Fluvioglacial terraces (Middle Pleistocene), 5) Pre-Alpine reliefs, 6) Main fluvial scarps, 7) Study area.

m. They consists of three basins (Lago Superiore, di Mezzo and Inferiore) at different elevation.

The surficial deposits of the study area have a grain size constituted mainly by gravel; however, sandy, clay-silt and peat deposits also crop out.

From the deep-geological point of view, the study area lies on the "Pedalpine Homocline" which represents the continuation, within the subsurface of the Po Plain, of the homocline structure that characterizes the Alpine margin in the Lake Garda area. This structure dips regularly southwards from the Alpine margin as far as the Po River (see Pieri & Groppi 1981).

The origin of the Mantova Lakes can be traced to the deviation of the River Mincio at Grazie with resulting abandonment of its course along a paleobed southwest of Mantova. According to Baraldi et al. (1980a), Zanferrari et al. (1982), Slejko et al. (1987) and Panizza et al. (1988), this diversion is connected to activity of a buried tectonic structure (Mantova Lakes Fault or Mantova Line). Different explanations were, however, given by previous authors (see Cozzaglio 1933, Veggiani 1974).

3 METHODS OF STUDY

The study was conducted according to the following investigation phases:

1. Bibliographic research. Various papers were reviewed concerned with the geomorphology, geology, surface hydrography, hydrogeology and human intervention in the sector of the Po Plain comprising the study area.

2. Analysis on topographical maps. The maps examined comprise the following:

a) Regional Technical Maps (C.T.R.) of the Lombardy Region (scale: 1:25,000 and 1:50,000) updated in 1983;

b) Istituto Geografico Militare Italiano (IGMI) maps (scale: 1:25,000), 1885, 1912, 1935, 1954 and 1973 editions of the F. 62 Mantova.

3. Morphological analysis through the interpretation of aerial photographs taken in various periods. The aerial photographs examined include black and white aerial photographs to a scale of 1:33,000 (approximately) taken in 1955 and to a scale of 1:70,000 (approximately) taken in 1994, and color aerial photographs taken in 1981 to a scale of 1:20,000.

4. Processing of altimetric data. On the basis of the altimetric data reported on the C.T.R. of the Lombardy Region, a "Microrelief map" was prepared, i.e. an altimetric map with 1 m equidistance. The altimetric map of the area was handly made by tracing the contour lines on the basis of the altitude reference points of the C.T.R. of the Lombardy Region. However, the altitude reference points corresponding to human artefacts (e.g. roads, bridges, etc.) were disregarded, and the contour lines in the quarrying areas and former quarrying areas were not traced since their altimetric situation does not tally with the original one. In the northernmost part of the study area greater graphic clarity required that only the main contours with 5 m range be given.

5. Field survey. The field survey was carried out in order to verify the present-day morphology and to collect data on the lithology of surface sediments. Surface deposits were studied by a simple standardised field procedure where over a thousand samples were collected at depths ranging from 0.5 to 1 metre (immediately below the level affected by pedogenesis and reworked by agricultural operations). About one hundred samples were subjected to laboratory analysis for precise classification. This resulted in a "Surface lithology map".

6. Elaboration of geomorphological maps. On the basis of the above investigations, two geomorphological maps were prepared: one referring to the features in 1997 and one referring to the features in 1955.

7. Geomorphological impact assessment. The geomorphological impact assessment of the study area was performed essentially on the basis of comparison between the morphological features represented in the maps as indicated in the above mentioned point and by applying a methodology already employed in other areas of Italy (e.g. see Panizza et al. 1996b). Because it is one of the most noticeable point in this study, it will be explained in deep in chapter 8.

4 MAIN HUMAN ACTIVITIES IN THE STUDY AREA

From the early 1960s onwards, the province of Mantova was subject to intense economic development. This resulted in an intensification of the effects of human action upon the natural environment. These resulted in profound changes in the geomorphological features of the Mincio plain, and can be grouped under three categories: a) quarrying, b) hydraulic management, and c) urban development.

4.1 *Quarrying activities*

The study area is characterised by extensive deposits of gravels from ground level to a depth of 10-15 m (Baraldi et al. 1976). The petrographic features of the gravels are such as to render them of high commercial value: carbonatic, magmatic and metamorphic rocks outcrop in various percentages. The deposits of inert material consist of lithoid elements generally well selected, with granulometric classes ranging from pebbles to sands (Fig. 3).

In the Roman period, the pebbly-gravelly deposits, easily detectable, were used in the construction of important roads. Quarrying activities at artisan level are documented in the Mincio plain from the early 20th century (Belenghi 1908): already at that time the Rivalta quarries not only supplied inert material for the needs of the province, but also produced large quantities for the neighbouring provinces. It was not till the early 1960s that quarrying became an industrial activity. From then on, there was a large and consistent increase in the amounts quarried (Fig. 4). Recent investigations (Baraldi & Zavatti 1990, Dal Ri 1991, Amministrazione Provinciale di Mantova 1996) have shown how the study area continues to supply more than 70% of the inert materials quarried in the province of Mantova, where more than 2,000,000 m³ are quarried per year.

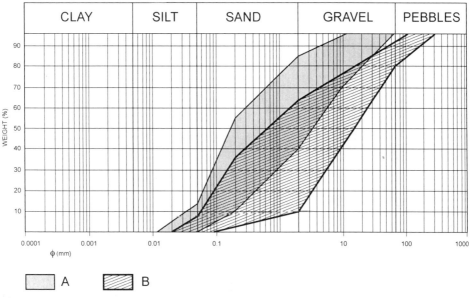

Figure 3. Range of the granulometry of the inert materials quarried in the Mincio River plain: A) Southern area between Goito and the Lakes of Mantova, B) Northern area between Pozzolo and Goito.

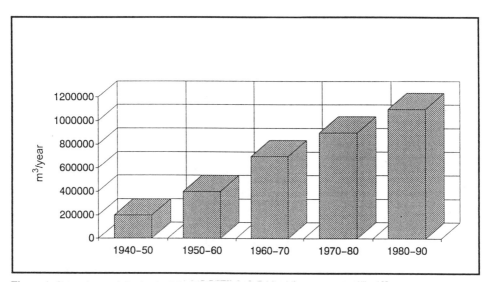

Figure 4. Quarrying activity in the Mincio River plain between 1940 and 1990.

The open quarries are classified as (Baraldi et al. 1980a): a) quarries exploited above the water table, b) trench quarries exploited below the water table (Fig. 5). The quarries of the first type are exploited only a few metres below the original ground level and in any case down to about 1 m above the level of maximum excursion of the water table. When no longer in use, these quarries are reclaimed for farming after laying a pedogenised level of organic soil on their floor (Fig. 6). To this type also belong the fluvial terrace quarries. These entail excavation starting from the edge of the terrace regression of the quarry front and consequent reduction or complete removal of the terrace itself. The quarries of the second type reach depths of up to 20 m, and when no longer in use are abandoned or equipped for recreational fishing or in some cases used as occasional dumping sites.

Figure 5. Trench quarry exploited below the water table.

Figure 6. Quarry exploited above the water table and reclaimed for farming practice after the laying of a pedogenised level of organic soil on the floor.

4.2 Hydraulic management

The study area is whitin the Lake Garda catchment area. The main stream is the Mincio which is characterised by high-water predominantly in autumn and spring, and minimum flows in winter. Its mean flow rate is registered at the outflow from Lake Garda (60-80 m^3/s, though this may vary from a minimum of 30 to a maximum of 200 m^3/s), while further down it falls considerably owing to the withdrawal of water for irrigation (Autori Vari, 1990). The flow rate of the Mincio is regulated by a weir a little south of Lake Garda.

In the Pozzolo stretch the Mincio has been canalised, which has altered its geometry and gradient, requiring several artificial steps. At Pozzolo, a regulating weir enables the Mincio to flow at 70 m^3/s, while in a leftwards direction flows the Scaricatore Pozzolo-Maglio (whose construction was begun in the early 1960s) which takes a maximum flow rate of 130 m^3/s. After a course of about 14 km, this artificial canal joins the Diversivo Mincio, a canal which is completely embanked and concreted whose construction started in the mid-1950s. The Diversivo Mincio, which diverges from the river to south of Goito, is responsible for collecting the flows of the Mincio that exceed 50 m^3/s. This is a 18 km canal which returns to the River Mincio bed to southeast of the Mantova lakes.

The water defences of the city of Mantova were the object of attention as early as the 12th century when the city was surrounded by lakes placed at different elevations. Other works for water defence were established during the 17th century (Bertazzolo 1609), in order to prevent the high water of the Po, returning along the course of the tributary Mincio, from flooding the city.

The Diversivo Mincio and the Scaricatore Pozzolo-Maglio solved the problem of Mantova's defence by water, but have had serious repercussions on the environment. For instance, as well as greatly altering the geomorphology of the study area, they have also brought about hydrogeological changes, causing the piezometric surface of the water table to fall over a vast radius around the main canals. In addition, they have hindered the occurrence of the high-water which formerly cleaned the Mantova lakes of the large mass of rotting vegetation and silts in suspension, and thus periodically helped to maintain the flora-fauna environment (Autori Vari 1990). This made it necessary to reclaim extensive areas on both shores of the lakes, which was in fact undertaken between the late 1960s and the early 1980s (Baraldi et al. 1990).

4.3 Urban development

From the early 1960s on, as a result of strong economic expansion and complex changes in the social structure of the family (Amministrazione Provinciale di Mantova 1984), there was a greater demand for new housing. Since this could not be met by the provincial capital itself, because of the presence of the lakes, it stimulated the development of small built-up areas on the periphery of Mantova. Thus, the built-up areas of the Mincio plain have undergone considerable expansion, with very evident increase in the area occupied by residential building, services and infrastructures, trading, artisanal and industrial areas. While certain urban areas have developed above all on the main level of the plain (S. Antonio, Bancole, Rivalta) others have also invaded the depression of the River Mincio (Pozzolo, Marmirolo, Goito, Soave), with irreversible alteration to the landscape.

5 PRESENT-DAY GEOMORPHOLOGY (YEAR 1997)

In the preparation of the maps described below, the Regional Technical Maps (C.T.R.) of the Lombardy Region and the aerial photographs taken in 1981 and in 1994 were used.

5.1 Processing of altimetric data

To define the altimetric set-up of the study area a "Microrelief map" was made, i.e. an altimetric map with 1 m equidistance (Fig. 7).

From analysis of the microrelief map it can clearly be seen that the study area is divided into two quite different sectors: the main level of the plain and the depression in which the Mincio flows.

The main level of the plain shows a general inclination roughly from north to south. The topographic extremes are represented by heights of 80 m at the north end of the area and

Figure 7. Microrelief map. 1) Contour lines (equidistance 1 m; equidistance 5 m in the northern sector), 2) Relative maximum and minimum altitudes. For the other symbols see Figure 10.

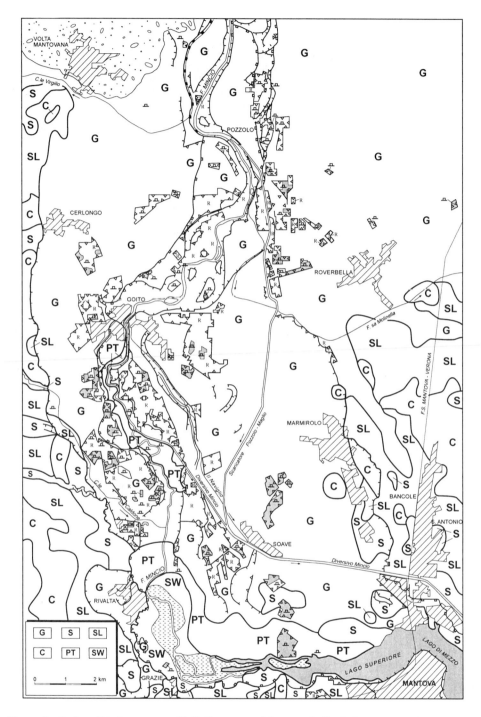

Figure 8. Surface lithology map. G) Sediments, mainly gravel, S) Sediments, mainly sand, SL) Sediments, mainly silt, C) Sediments, mainly clay, P) Peat deposits, SW) Swampy deposits. For the other symbols see Figure 10.

about 28 m in the southern sector. The overall acclivity of the territory is around 0.3%. However the acclivity is not uniform, and the area can be further divided into two sub-sectors north and south of the Cerlongo-Roverbella alignment. In the north sub-sector, two alluvial fans can be detected. This zone has an acclivity of around 0.5%; the south sub-sector as a lesser acclivity of about 0.1%. Within the Mincio depression, the altimetric picture is complicated not only by several quarries but also by scarps of various height mainly in a N-S direction, which give rise to various orders of terrace. In this sector, too, a general inclination from north to south can be detected; the altimetric maxima are represented by heights of about 60 m at the northern apex of the depression itself and about 17 m at the Lake Superiore of Mantova; the overall acclivity of this territory is around 0.28%.

5.2 *Lithology of the surface sediments*

The distribution of the prevalent textures in the superficial sediments is represented in a "Surface lithology map" (Fig. 8).

The most widespread deposits in the study area are mainly gravel sediments, but always mixed with a certain percentage of sand. These outcrop continuously from the edge of the moraines as far as the southern sector. Both the fluvioglacial deposits of the outwash plain and the fluvial deposits of the Mincio depression belong to this class. They are practically impossible to differentiate owing to similar deposition processes and the identical petrographic composition of the clasts. In the gravelly deposits, observed in the numerous quarries, well-rounded clasts of differing granulometry are found. The stratification is more or less regular or lenticular, sometimes crossed or truncated.

The mainly sand deposits outcrop essentially on the main level of the plain: in small zones at the western edge of the study area and, east of the Mincio depression, in the area of S. Antonio. Some stretches also occur within the Mincio depression near the Mantova lakes. From the textural point of view the mainly sand deposits show a considerable variety (sand percentages from 40% to 80%).

The mainly silt and mainly clay sediments are present in the same sectors where the sands outcrop. The first generally outcrop in areas at the sides of current and extinct water courses. They display a mixed granulometry in which the silty fraction varies between 40% and 60%. The mainly clay sediments are often found dovetailed with the mainly silt and mainly sand sediments. They are rich in $CaCO_3$, which may be present in the form of irregular concretions (diameter from 0.5 to 100 mm) or of extensive crusts located in the zone of fluctuation of the water table. The granulometry of the mainly clay sediments also displays a considerable variety (clay from 45% to 70%) and is always accompanied by important fractions of silt.

The peat deposits, which generally cover mainly gravel deposits, appear south of Goito, in a small stretch along the sides of the river Mincio (connected with abandoned meanders and zones of water stagnation), and on the left shore of the Lake Superiore of Mantova (here they attain thicknesses ranging from 30 to 100 cm). Their presence is, however, diminishing as a result of human intervention.

Lastly, in the area where the Mincio enters the Lake Superiore of Mantova swampy deposits are found, consisting of vegetable residues mixed with variable silt and clay fractions. Human intervention is also causing these deposits to diminish.

5.3 *Geomorphological map*

The geomorphological map (Fig. 9) was prepared by referring to the official legend of the "Geomorphological Map of the Po Plain" project (Castiglioni et al. 1986) and its subsequent modifications (Fig. 10). In the northwestern zone of the study area the morainic hills of the Lake Garda amphitheatre (Upper Pleistocene) outcrop. As said before, the plain in front of the morainic hills corresponds to the outwash plain of the Upper Pleistocene. The northern part of the outwash plain is characterised by alluvial fans (notable among these is the large fan that develops to the north of Roverbella) and by widespread traces of braided streams. Further down, the outwash plain is furrowed by a system of paleobeds with direction varying from N-S to WNW-ESE. The outwash plain is deeply incised by the River Mincio which has given rise to a wide triangle-shaped depression flanked by fluvial scarps a few metres high.

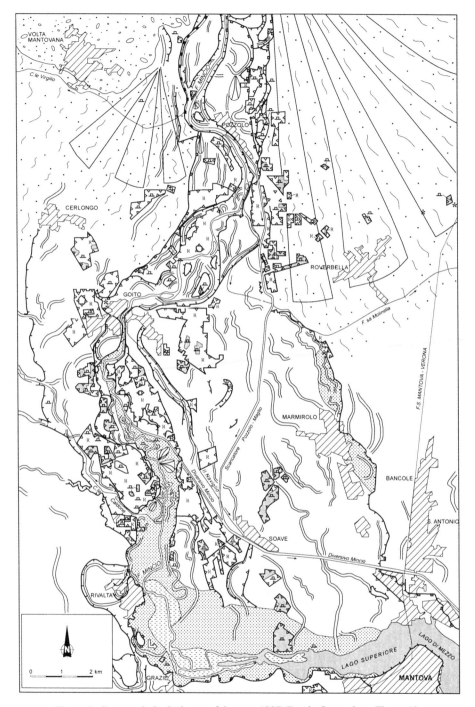

Figure 9. Geomorphological map of the year 1997. For the Legend see Figure 10.

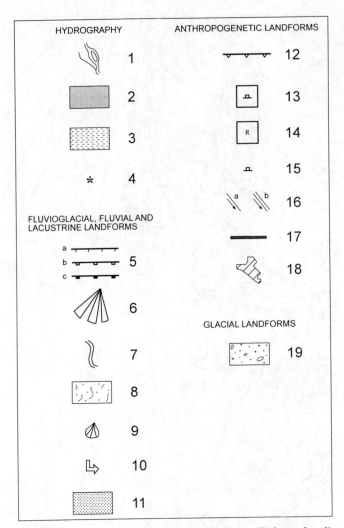

Figure 10. Legend of the Geomorphological Maps. Hydrography: 1) River bed, canal, 2) Lake, pond, 3) Swamp, 4) Resurgence. Fluvioglacial, fluvial and lacustrine landforms: 5) Terrace scarp or edge of fluvial scarp (a: less than 5m high; b: 5 - 10 m high; c: more than 10 m high), 6) Alluvial fan, 7) Trace of paleobed, 8) Widespread traces of abandoned braided streams, 9) Crevasse splay, 10) Site of river diversion, 11) Reclaimed swampy area. Anthropogenetic landforms: 12) Edge of artificial scarp, 13) Quarrying area, 14) Quarry reclaimed to farming practice, 15) Small quarry, 16) Main artificial canal: a) without levees; b) with levees, 17) Stretch of canal under construction, 18) Main built-up areas. Glacial landforms: 19) Morainic hills of the amphitheatre of Lake Garda (Upper Pleistocene).

This depression features various orders of fluvial terraces. Five terraces have been identified, in the northern zone near Pozzolo, besides the present Mincio plain (see Castaldini 1994) and three in the southern zone near the Mantova Lakes. The scarps bounding the various orders of terrace diminish in height with distance downstream (from heights of over 10 m to heights of a few metres) giving rise to a situation of converging terraces.

A particular feature, left of the Mincio, is the modest scarp (1-2 m high) that develops, in a NNW-SSE direction, between Cerlongo and Rivalta. This form may be ascribed to a paleo-

Mincio flowing on the outwash plain before it began to cut the depression in which it currently flows.

Along its course the Mincio has the appearance of a modest-sized channel and flows in very varying morphological situations. In its northernmost stretch, its course is within different orders of fluvial terrace. In its southern stretch as far as the Mantova Lakes it flows at the level of the surrounding plain. In the central stretch it follows a sinuous course with fairly regular meanders. In the depression, a large number of paleobeds of the Mincio are found, among them various meander cut-offs near the present course. Some crevasse splays have also been found near both the present course and the most recent paleobeds.

Currently carrying only pelitic sediments, the Mincio conveys its waters into the Lake Superiore of Mantova after flowing the last stretch within a swamp where it anastmoses. The geomorphological map also shows the reclaimed swampy areas and the important diversion of the Mincio in the vicinity of Grazie. The natural hydrology also features the "resurgences", important springs which in the past have led many minor water courses. Those currently found (in the zones of Roverbella and Goito) represent only a small part of those formerly existing (see Baraldi & Pellegrini 1978).

Anthropogenetic activities have played an important role in the morphogenesis of the study area. As regards the main artificial canals, the geomorphological map details the Scaricatore Pozzolo-Maglio and the Diversivo Mincio, and their functions were described above. With reference to the urbanised areas in the geomorphological map, only the main built-up areas are shown (this in order not to crowd out the natural elements). The most extensive urban centres are Bancole-S. Antonio, Marmirolo and Goito.

Quarrying activity caused a great deal of damage to the landscape. The quarries exploited above the water table are present as large, deep, regular depressions in the terrain. In the areas of quarry reclaimed for farming, where excavation was limited to the first few metres of the ground level, one may note small groups of houses perched on the top of isolated artificial reliefs and the electricity pylons aligned on tiny mounds of mainly gravelly soil. The trench quarries below the water table have formed several small ponds which are extremely different in shape from natural internal waters. The quarrying areas occur more or less everywhere, but concentration occurs in the strip of territory along the sides of the River Mincio, both on the outwash plain and in the depression of the river.

6 MORPHOLOGY IN THE YEAR 1955

The morphological situation in 1955, shown in Figure 11, has been derived from analysis of the 1954 edition of the IGMI maps and from the aerial photographs taken in 1955.

For this period elaboration of a contour map based on the altitude reference points of the IGMI maps was felt to be inappropriate, since these altitude reference points are few and generally corresponding to human artefacts located at heights above the natural ground level. So, it would not be comparable with the altimetric map based on the altitude reference points of the C.T.R. (Fig. 7).

From the aerial photographs the outwash plain can clearly be seen, cut through by the River Mincio. Within the triangle-shaped depression the various orders of terraces, can easily be detected. They appear practically intact. On the surface of these terraces can be seen numerous paleobeds at places some kilometres long and, at the sides of the present Mincio, several meander cut-offs and some crevasse splays. Close to the Mantova Lakes the active and reclaimed swampy zones can be clearly distinguished. Besides the resurgences shown in the 1997 Geomorphological map, additional areas were found in the sector between Goito and Volta Mantovana. The human intervention is fairly limited. Indeed, only a score of quarries are found, of modest extension, concentrated in the Goito area. In general, these are dry-floor quarries. In addition, to the northwest of Soave and south of S. Antonio, there appear some stretches of quarrying for the construction of the Diversivo Mincio. The built-up areas, too, are of limited extent.

7 COMPARISON BETWEEN FEATURES IN 1997 AND IN 1955

Comparison of the features represented in the geomorphological maps of 1997 (Fig. 9) and of 1955 (Fig. 11) shows how the most marked differences can be ascribed to human intervention and, especially, to quarrying activities. The latter is well known to pose

Figure 11. Geomorphological map of the year 1955. For the Legend see Figure 10.

problems of environmental impact in general, whose complexity directly depends on the geological features of the sites, the extension and depth of the quarries and the technical means of excavation and environmental recovery (Autori Vari 1982, Dal Ri 1991). In the case of trench quarries exploited below the water table, alterations in the piezometric level and the direction of flow of the ground water may occur (Avanzini & Beretta 1992). The interception at ground level of the most superficial aquifer causes potential pollution hazards owing to the possible uncontrolled inflow of waste materials. In addition, there may be impermeabilisation of the bottom or walls of the trench, leading to ground water buffering.

Where the quarries are exploited above the water table, the effects are the alterations to the pedologic characteristics, and hence the fertility of the topsoils, which can seldom be remedied by subsequent restoration work. The quarries exploited in the fluvial terraces causes the partial or total destruction of these relict landforms. Another effect related to the stability of the artificial scarps, often shaped without taking account of the geotechnical stability parameters. These slopes, as they are no longer stabilised even by the vegetation cover, are subject to rill wash, leading to erosion and widespread instability. The main effects and environmental changes caused by quarrying in the Mincio plain are summarised in Table 1.

Table 1. Main environmental modifications caused by quarrying activity. A) Quarrying above the water table, B) Quarrying below the water table.

	Type of modification	A	B
	1 - creation of regular - shaped depressions	X	X
	2 - creation of small isolated artificial reliefs	X	
GEOMORPHOLOGY	3 - creation of artificial ponds		X
	4 - partial or total destruction of fluvial terraces	X	
	5 - erosion and instability of quarry scarps	X	
	1 - depression of the piezometric surface		X
HYDROGEOLOGY	2 - alteration of ground water flow direction		X
	3 - formation of periodically flooded areas	X	X
	1 - permanent removal of areas from farming use		X
SOIL USE	2 - alteration of pedological characteristics of soil	X	
	3 - alteration of farming practices	X	

In more detail it can be seen how quarrying has led to the following changes in the natural landcape: a) partial destruction of various orders of terrace from first to fourth orders in the Pozzolo area; b) total destruction of various orders of terrace both right and left of the River Mincio between Goito and Rivalta; c) creation of vast, sometimes deep depressions between Pozzolo and Rivalta. These areas are nowdays below the natural level of the plain, and may flood in periods when the water table is high; and d) creation of very numerous geometric-shaped ponds between Pozzolo and the Lake Superiore of Mantova.

As regards urban development, comparison between the geomorphological maps reveals a considerable expansion of the smaller urban centres (especially Goito and Marmirolo). Mantova, whose expansion is conditioned by the presence of its lakes, shows no substantial differences. The urban development of Mantova has, however, been clearly oriented towards the main level of the plain northwards, giving rise to a vast built-up area that has absorbed S. Antonio and Bancole, which in 1955 were separate.

Concerning the hydrology, the course of the Mincio shows no significant changes, except for the zone where it flows into the Lake Superiore of Mantova. In this zone the 1955 configuration of the various branches appears more complex as compared to the present situation. Also there are marked differences as regards the extent of the Mantova Lakes and their swamps. In 1955 the area of the lakes was smaller and that of the swamps larger than at present. This is connected with the fact that between the late 1960s and the early 1980s large swampy areas were reclaimed (Baraldi et al. 1980c). Serious alterations to the landscape and the hydrogeological situation of the water table have also been caused by the hydraulic works. In particular, the excavation of the Scaricatore Pozzolo-Maglio and the Diversivo Mincio have entailed moving some two million cubic metres of inert material (gravels, sands and silts), the interruption of the morphological continuity of the sites, and a different hydraulic situation even in the minor water courses. Moreover lining the above-mentioned canals with concrete facing has hindered the natural flow of ground waters, causing important variations in the piezometric level of the surface aquifer, with diminution of up to about 5 metres.

Comparison of the 1955 hydrogeological data with the current ones shows alterations that can be retraced to two different typologies (Fig. 12): a) area where the hydraulic works have intercepted the water table, changing its flow from N-S to NNW-SSE, and b) area where quarrying has caused a drop in the water table of about one metre, as well as an area where the hydraulic work has caused a piezometric fall in the water table of about 5 metres.

The lowering of the water table is also emphasided by the riduction of the resurgences.

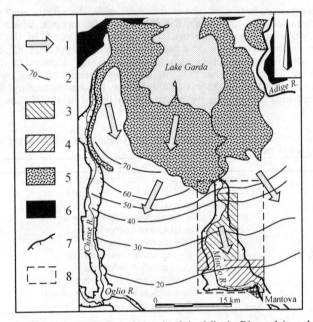

Figure 12. Hydrogeological sketch of the Mincio River plain and its modifications between 1955 and 1997. 1) Axis of underground drainage, 2) Piezometric isolines and their altitude (metres a.s.l.), 3) Area in which the direction of flow of the ground waters has changed, 4) Area where there has been a piezometric fall in the water table, 5) Morainic hills (Middle-Upper Pleistocene), 6) Pre-Alpine reliefs, 7) Edge of main fluvial scarp, 8) Study area.

8 GEOMORPHOLOGICAL IMPACT ASSESSMENT

On the basis of the geomorphological maps, the geomorphological impact on the study area was assessed with a semiquantitative procedure. A landform can be considered as a geomorphologic asset. It can be regarded as such not only on the basis of scenic, social, economic and cultural value, but also on the basis of criteria related to the more general scientific concept of an asset. First the geomorphological assets of the area were identified and evaluated. Then the scientific quality and the impact were determined by applying a simplified methodology (Panizza 1996, Panizza et al. 1996b) set out for European Union projects (Marchetti et al. 1995, Panizza et al. 1996a).

All the landforms of the landscape in the 1955 geomorphological map were classified on the basis of four different characters, disregarding cultural, socioeconomic, aesthetic and other attributes (Carton et al. 1994) which do not directly concern the geomorphological study, but have to do with other specific fields of study.

Each of the landforms in the 1955 geomorphological map was then transformed in a polygon (Fig. 13). Each of these polygons (N of Tab. 2) can be classified by a two-figure code (F) which enables description of the landforms (e.g. fluvial scarps) and surfaces characterised by non-distinguished deposits (e.g. outwash plain).

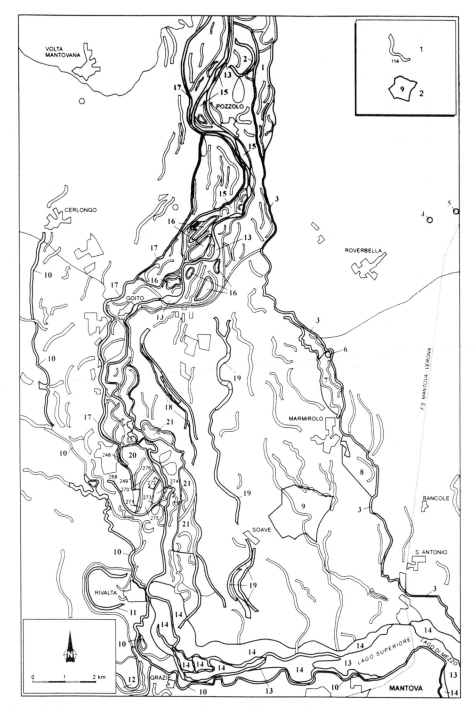

Figure 13. Geomorphological asset map of the year 1955. 1) Landform polygon and its order number, 2) Geomorphological asset and its order number. The singol landform polygons are numbered, as an example, only within the asset n. 20.

Table 2. Geomorphological Asset table. N) Number of the polygon; F) Geomorphological classification of the polygon; G,O,P,E) Characters of the polygon (G = model of geomorphological evolution; O = object used for educational purposes; P =paleogeomorphological example; E = ecological support); A) Order number of the Geomorphological Asset containing the polygons; V) Scientific Value; C) Degree of Preservation; Q) Scientific Quality; D) Geomorphological Deterioration; I) Geomorphological Impact.

N	F	G	O	P	E	A	V	C	Q	D	I
9	27	0.25	0.5	0.5	0	17	1.25	2	2.5	1	0
18	28	0.25	0.25	0	0	15	0.50	1.5	0.75	1	0
19	28	0.25	0.25	0	0	15	0.50	1.5	0.75	1	0
29	26	0.25	0.5	0.5	0	17	1.25	2	2.5	0.75	0.625
42	34	0.25	0.25	0	0	16	0.50	1.5	0.75	0.25	0.5625
44	28	0.25	0.25	0	0	15	0.50	1.5	0.75	0.25	0.5625
46	34	0.25	0.25	0	0	16	0.50	1.5	0.75	0	0.75
47	25	0.25	0.25	0	0	16	0.50	1.5	0.75	0	0.75
48	34	0.25	0.25	0	0	16	0.50	1.5	0.75	0	0.75
49	34	0.25	0.25	0	0	16	0.50	1.5	0.75	0	0.75
50	25	0.25	0.25	0	0	16	0.50	1.5	0.75	0	0.75
51	25	0.25	0.25	0	0	16	0.50	1.5	0.75	0	0.75
52	21	0.25	0.25	0	0.75	13	1.25	1.5	1.875	0.75	0.4687
53	34	0.25	0.25	0	0	16	0.50	1.5	0.75	1	0
54	25	0.25	0.25	0	0	16	0.50	1.5	0.75	1	0
60	34	0.25	0.25	0	0	15	0.50	1.5	0.75	0.75	0.1875
61	28	0.25	0.25	0	0	15	0.50	1.5	0.75	1	0
62	28	0.25	0	0	0	15	0.25	1.5	0.375	1	0
63	28	0.25	0.25	0	0	15	0.50	1.5	0.75	1	0
67	34	0.25	0.25	0	0	15	0.50	1.5	0.75	1	0
68	34	0.25	0.25	0	0	15	0.50	1.5	0.75	1	0
69	34	0.25	0.25	0	0	15	0.50	1.5	0.75	1	0
70	34	0.25	0.25	0	0	15	0.50	2	1	1	0
71	34	0.25	0.25	0	0	15	0.50	2	1	0.75	0.25
72	28	0.25	0	0	0	15	0.25	1.5	0.375	1	0
74	25	0.25	0.25	0	0	15	0.50	1.5	0.75	1	0
77	28	0.25	0.25	0	0	15	0.50	1.5	0.75	1	0
78	28	0.25	0.25	0.25	0	2	0.75	2	1.5	1	0
84	26	0.25	0.5	0.5	0	3	1.25	2	2.5	0.9	0.25
85	28	0	0	0.25	0	1	0.25	1	0.25	0	0.25
86	33	0.25	0.25	0.25	0	1	0.75	2	1.5	0.5	0.75
87	25	0.25	0.25	0.25	0	1	0.75	2	1.5	0.75	0.375
88	25	0.25	0.25	0.25	0	1	0.75	2	1.5	1	0
89	25	0.25	0.25	0	0	1	0.50	2	1	1	0
90	33	0.25	0	0.25	0	1	0.50	2	1	1	0
91	33	0.25	0	0.25	0	1	0.50	2	1	1	0
92	33	0.25	0	0.25	0	1	0.50	2	1	1	0
93	25	0.25	0.25	0	0	1	0.50	2	1	1	0
94	33	0.25	0	0.25	0	1	0.50	1.5	0.75	1	0
95	28	0	0	0.25	0	1	0.25	1	0.25	0.25	0.1875
96	28	0	0	0.25	0	1	0.25	1.5	0.375	1	0
97	25	0.25	0.25	0.25	0	1	0.75	1.5	1.125	1	0
98	28	0	0	0.25	0	1	0.25	1	0.25	0	0.25
99	33	0.25	0.25	0.25	0	1	0.75	2	1.5	0.5	0.75
101	24	0	0.5	0	0.25	4	0.75	1.5	1.125	1	0
102	24	0	0.5	0	0.25	5	0.75	1.5	1.125	1	0
105	25	0.25	0.5	0.5	0	3	1.25	1.5	1.875	0.75	0.4687
117	34	0.25	0.25	0	0	16	0.50	1.5	0.75	1	0
118	34	0.25	0.25	0	0	1	0.50	1.5	0.75	0.9	0.075
120	25	0.25	0.25	0	0	16	0.50	1.5	0.75	1	0
121	25	0.25	0.25	0	0	16	0.50	1.5	0.75	1	0
134	28	0	0.25	0	0	19	0.25	2	0.5	1	0
135	25	0	0.25	0	0	19	0.25	2	0.5	0.9	0.05
137	25	0	0.25	0.25	0	10	0.50	1.5	0.75	1	0
149	25	0	0.25	0.25	0	10	0.50	1.5	0.75	1	0
154	25	0	0.25	0.25	0	10	0.50	2	1	1	0
156	25	0	0.25	0.25	0	10	0.50	2	1	0.9	0.1
175	28	0.25	0.25	0	0	18	0.50	1.5	0.75	1	0
176	25	0.25	0.25	0	0	18	0.50	2	1	1	0
186	28	0.25	0.25	0	0	18	0.50	1.5	0.75	1	0
187	28	0	0.25	0	0	19	0.25	2	0.5	0.9	0.05

Table 2 continued

N	F	G	O	P	E	A	V	C	Q	D	I
203	28	0	0.25	0.25	0	7	0.50	1.5	0.75	1	0
204	28	0	0.25	0.25	0	7	0.50	1	0.5	1	0
205	24	0	0.25	0	0.25	6	0.50	1.5	0.75	1	0
207	28	0	0.25	0	0	7	0.25	1.5	0.375	1	0
208	30	0.25	0.25	0	0.25	7	0.75	1.5	1.125	1	0
215	30	0.25	0	0	0.25	8	0.50	1	0.5	0.75	0.125
216	28	0.25	0	0	0	8	0.25	1	0.25	1	0
221	26	0.5	0.5	0.5	0	3	1.50	1.5	2.25	0.5	1.125
223	23	0.25	0	0	0.5	14	0.75	1.5	1.125	0.75	0.2812
233	23	0.25	0	0	0.5	14	0.75	1.5	1.125	0	1.125
237	28	0.25	0.25	0	0	19	0.50	2	1	1	0
238	25	0.25	0.25	0	0	19	0.50	2	1	0.9	0.1
239	25	0.25	0.25	0	0	19	0.50	2	1	0.75	0.25
249	28	0.25	0	0	0	20	0.25	1	0.25	0	0.25
258	25	0.25	0.25	0	0	21	0.50	1.5	0.75	0.25	0.5625
261	33	0.25	0.25	0	0	21	0.50	1.5	0.75	0.25	0.5625
262	25	0.25	0.25	0	0	21	0.50	1.5	0.75	0.25	0.5625
263	28	0.25	0.25	0	0	21	0.50	1.5	0.75	0	0.75
268	25	0.25	0	0	0	20	0.25	1.5	0.375	0.5	0.1875
270	25	0.25	0.25	0	0	20	0.50	1	0.5	0	0.5
271	28	0.25	0	0	0	20	0.25	1	0.25	0	0.25
273	25	0.25	0.25	0	0	20	0.50	1.5	0.75	0	0.75
274	34	0.25	0.25	0	0	20	0.50	1.5	0.75	0.25	0.5625
275	33	0.25	0.25	0	0	20	0.50	1.5	0.75	0	0.75
276	33	0.25	0.25	0	0	20	0.50	1.5	0.75	0.1	0.675
292	26	0.25	0.5	0.25	0	10	1.00	2	2	0.9	0.2
293	25	0	0.5	0.5	0	12	1.00	2	2	1	0
294	28	0	0	0.25	0	11	0.25	1.5	0.375	1	0
306	23	0.25	0.25	0	0.75	14	1.25	1.5	1.875	1	0
307	23	0.25	0	0	0.75	14	1.00	1.5	1.5	1	0
308	23	0.25	0	0	0.75	14	1.00	1.5	1.5	1	0
309	23	0.25	0	0	0.75	14	1.00	1.5	1.5	0.75	0.375
310	23	0.25	0	0	0.75	14	1.00	1.5	1.5	0.75	0.375
313	25	0	0.25	0.5	0	12	0.75	1.5	1.125	1	0
314	28	0	0.5	0.5	0	12	1.00	2	2	1	0
317	33	0	0.25	0.5	0	12	0.75	1	0.75	1	0
318	33	0	0.25	0.5	0	12	0.75	1.5	1.125	1	0
321	25	0	0.25	0.25	0	10	0.50	1	0.5	0.75	0.125
322	23	0.25	0	0	0.25	14	0.50	1	0.5	0	0.5
324	25	0.25	0.25	0.25	0	21	0.75	1.5	1.125	0.5	0.5625
325	28	0.25	0.25	0.25	0	21	0.75	1.5	1.125	0.9	0.1125
331	23	0.25	0.25	0	0.75	14	1.25	2	2.5	0.25	1.875
334	28	0.25	0	0	0	21	0.25	1.5	0.375	1	0
335	28	0.25	0.25	0	0	21	0.50	1.5	0.75	1	0
336	34	0.25	0.25	0	0	21	0.50	1.5	0.75	0.5	0.375
337	25	0.25	0.25	0	0	21	0.50	1.5	0.75	0	0.75
338	25	0.25	0	0	0	21	0.25	1.5	0.375	0	0.375
339	34	0.25	0.25	0	0	21	0.50	1.5	0.75	0.25	0.5625
340	34	0.25	0	0	0	1	0.25	1.5	0.375	0.75	0.0937
343	33	0.25	0	0.25	0	1	0.50	1.5	0.75	1	0
345	28	0	0	0	0.25	9	0.25	1.5	0.375	1	0
346	32	0	0	0	0.75	9	0.75	2	1.5	1	0
347	32	0	0	0	0.75	9	0.75	2	1.5	1	0

The four characters that determine the geomorphological value of the polygon (Panizza 1996) are (see Tab. 2): a) model of geomorphological evolution (G); b) object used for educational purposes (O); c) paleogeomorphological example (P); and d) ecological support (E).

To each of these characters was assigned a level of interest varying between 0 and 1: non-significant (0), local (0.25), regional (0.5), super-regional (0.75), world-wide (1).

It should be pointed out that this semiquantitative method is affected by a considerable degree of subjectivity; nevertheless it is an attempt for codifying geomorphological parmeters for the quantification of the scientific quality of geomorphological assets. The rates thus obtained are not comparable with different areas of investigations and should, in any case, be interpreted by expert in this field.

The polygons selected for any one of the four characters can be classified as belonging to a geomorphological asset. Thus Figure 13 shows 21 geomorphological assets identified, which generally consist of several polygons (A of Tab. 2).

Assets 1, 20, 21 represent area of considerable geomorphological interest since they testify to the various stages of morphogenesis of a fluvial plain subject to terracing. The various orders of terrace are, moreover, well evident and of educational interest.

Assets 2, 7, 8, 11, 12, 18, 19 represent traces of abandoned meanders or , more generally, paleobeds. They are of importance in reconstructing late-quaternary hydrological variations and have great educational value.

Assets 3, 10, 17 represent the main scarps that divide the outwash plain from the depression of the River Mincio, and therefore are of great paleogeomorphological value and, in view of the evidence they provide, are also of educational importance.

Assets 4, 5, 6 are well preserved resurgences and are important for ecological support and their educational value.

Asset 9 represents one of the few remnants of plain forest (Bosco della Fontana) which should provide the natural arboreal covering for the whole Po Plain. Within the Bosco della Fontana much can be learnt from the modelling processes of the surface waters, which are influenced by the constant presence of a calcic horizon at less than 1 metre depth.

Assets 13, 14, 15, 16 are the most recent areas of the Mincio depression, and are important because they represent the ecological support necessary for the proliferation of a large number of animal and vegetable species typical of these wetlands. Such areas are extremely delicate, in unstable equilibrium, and protected by a regional park authority set up in 1984 (Natural Park of the River Mincio).

In order to assess the quality of the individual polygons, their degree of preservation was evaluated, so that the Quality (Q, Tab. 2) can be expressed as the product of the Value (V is the sum of the individual scores given to the characters of each polygon) for the degree of Preservation (C). This last can be expressed by a numerical index that in this instance was chosen between 1 and 2 (1 = poorly preserved; 1.5 = averagely preserved; and 2 = well preserved). In view of the general nature of this investigation, it was decided not to assign different weights to the four different characters. Obviously, different priorities could be assigned to the characters, with scores of different ranges for more specific researches, or more expanded ranges also for the degree of preservation.

Comparison between the 1955 and 1997 geomorphological maps has highlighted considerable variations. In particular, the increase of areas involved by human activity was examined. The differences emerging from comparison of the two geomorphological maps are shown in Figure 14. Comparison between the areas affected by human modification and the geomorphological assets selected enables assessment of the resulting deterioration of the geomorphological assets. The Deterioration (D in Tab. 2) was quantified with values going from 0 (characteristics of the landform destroyed) to 1 (no damage). For each landform classified as part of a geomorphological asset (A) we then calculated the Impact (I) undergone by each landform obtained as the product of the quality (Q) by the difference between unity and Deterioration (D). In Table 2 it can be seen how the greatest impacts in numerical terms are those occurring on polygons 331 e 221, i.e. the polygons of high quality on which a deterioration has occurred from high (331) to medium (221) and polygon 233 of medium quality which has been completely destroyed. In Figure 14 it can be observed how the impacts most significant from the area point of view are located in the main quarrying centres (assets 1, 20, 21, 15, 16) and along the left bank of the Mincio where the river widens to form the Lake Superiore of Mantova (asset 14). If one considers the total Quality of the geomorphological assets studied, it can be noted how asset 20 is the one that has undergone the greatest Impact, equivalent to 90% , while the impact on assets 21 and 16 has caused a loss of Quality of around 50% (Tab. 3).

Table 3. Percentage of Impact (% I) on the Geomorphological Assets (A).

A	1	2	3	4	5	6	7	8	9	10	11	12	13	14	15	16	17	18	19	20	21
%I	17	0	27	0	0	0	0	16	0	7	0	0	24	35	9	56	12	0	11	90	57

In terms of absolute values, no comparison is possible between the Quality of the individual geomorphological assets. For this methodology enables only the assessment of the Impact on the individual geomorphological asset, and does not allow comparison between them or assessment of absolute quality. In order to assess these further variables, other parameters would have to be included.

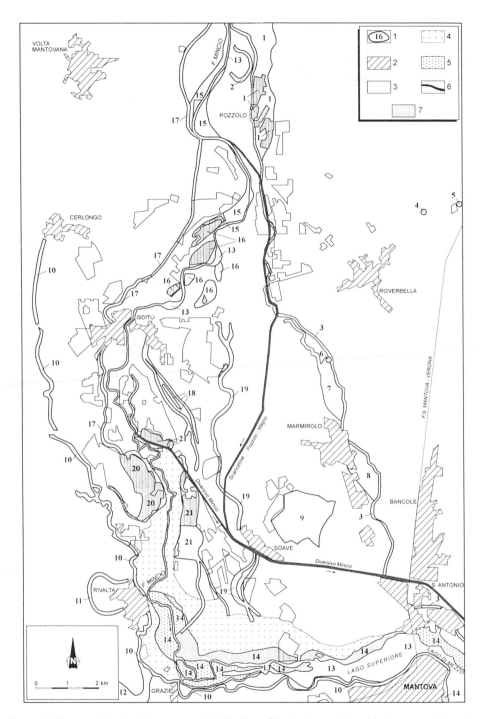

Figure 14. Geomorphological Impact map. Modification of the landscape induced by human activity after the year 1955. 1) Geomorphological Asset and and its order number (1955), 2) Main built-up areas (1997), 3) Quarrying area (1997), 4) Reclaimed swampy area (1997), 5) Swamp (1997), 6) Main artificial canal with levees, 7) Impact on Geomorphological Asset.

9 CONCLUSIONS AND REMARKS ON ENVIRONMENTAL RESTORATION

This study, aimed at assessing the geomorphological impact by comparing the features in the mid-1950s and that of today, has enabled the effect of human activity on the environment to be evaluated. The study shows that among the main human activities (quarrying of inert materials, hydraulic management and urban development) the most serious damage to the landscape has been caused by quarrying. The greatest impact, with over 50% of the quality of the geomorphological assets lost, has occurred in the vicinity of the River Mincio between Goito and Rivalta. In this area, quarrying has been responsible for a general lowering of the natural plain level with destruction of several orders of fluvial terraces and repercussions on the hydrogeological situation (variations of the piezometric level and directions of flow of ground waters). Serious alterations to the landscape and the hydrogeological situation of the water table have also been caused by the hydraulic works, in particular, the excavation of the Scaricatore Pozzolo-Maglio and of the Diversivo Mincio.

The intense quarrying activity in the Mincio plain has posed a number of problems for the local administration concerning the protection and recovery of the environment. Part of the Mincio plain is currently destined to become a Natural Fluvial Park. The park authority has recently set up very careful regulations for newly established quarrying firms, as well as guidelines for the environmental recovery of disused quarries. In particular, flooded-floor quarries are to undergo renaturalisation with the aim of devoting them to recreational or naturalistic activities. The dry-floor quarry will be partly earmarked for reafforestation, with trees typical of the area. It will be less easy to restore the natural fertility of the soils where the quarries are destined to be returned to agricultural use. Ongoing experience has shown that a long time will be needed in order to achieve acceptable vegetable productivity. In addition, owing to the altered hydrogeological conditions, in particular periods when the water table receives abundant supplies from rain or irrigation, ground water emerges in the quarried areas where no safe distance has been maintained between the new topographical surface and the piezometric level.

The main hydraulic works, undertaken according to criteria that do not aimed to protect the landscape, are under study by the Mincio Park authority, with the aim of placing them in the environmental context more effectively. Hence, there are ongoing bioengineering experiments targeted towards the restoration of strips of riparian vegetation along the canals, which would have the function of animal repopulation and antipollution barrier.

As regards the urban development, it should be underlined how, in the present situation, there is a complete absence of any territorial planning coordinated among the various levels of local government. That leads to a proliferation of service infrastructures that are gradually, but irreversibly, removing vast areas from the natural environment.

AKNOWLEDGEMENTS

The financial support for this study was provided by the Ministero dell'Università e della Ricerca Scientifica e Tecnologica (MURST) grants 40% (Coordinator of the Modena University, prof. Mario Panizza).

REFERENCES

Amministrazione Provinciale di Mantova 1984. *Popolazione e movimento anagrafico nei comuni mantovani*. Mantova: Assessorato Ecologia e Ambiente.
Amministrazione Provinciale di Mantova 1996. *Revisione del piano cave provinciale*. Mantova: Assessorato Programmazione,
Autori Vari 1982. *Le etudes d'impact de carrières de amtériaux alluvionaires*. Paris: Laboratoire central des Ponts et chausses.
Autori vari 1990. *Il Mincio e il suo territorio*. Verona: Cierre Edizioni.
Avanzini, M. & Beretta, G.P. 1992. Il sistema idrodinamico Falda-Cava in relazione alla struttura idrogeologica. *Proceedings of the first Conv. Giov. Ric. Geol. Appl.* 22-23 October 1991, Gargnano (Bs): 19-27.
Baraldi, F. & Pellegrini, M. 1978. I fontanili della pianura compresa tra i fiumi Chiese e Mincio (Province di Brescia e Mantova). *Quad. IRSA* 34(18): 439-446.

Baraldi, F. & Zavatti, A. 1990. Carta di vulnerabilità degli acquiferi a media scala: II. La provincia di Mantova. In A. Zavatti (ed), *Studi sulla vulnerabilità degli acquiferi*. 83-94. Bologna: Pitagora.

Baraldi, F. & Zavatti, A. 1994. Idrogeologia e idrochimica degli acquiferi in provincia di Mantova. In F. Baraldi & A. Zavatti (eds), *Studi sulla vulnerabilità degli acquiferi*: 1-32. Bologna: Pitagora.

Baraldi, F., Cantoni, A. & Novellini, G. 1976. *Le ghiaie della pianura mantovana tra Pozzolo, Roverbella, Marengo, Marmirolo, Soave, Rodigo e Goito*. Parma: Il Frantoio.

Baraldi, F., Cantoni, A. & Novellini, G. 1990. Caratteristiche geotecniche del sottosuolo della città di Mantova. *Proceedings of the VII Congr. Naz. O.N.G.*, Roma: 12-19.

Baraldi, F., Cantoni, A., Novellini, G. & Malago', R. 1980a. Censimento delle attività etrattive nella provincia di Mantova. *Acqua e Aria*, 6: 67-82.

Baraldi, F., Carton, A., Castaldini, D., Panizza, M., Pellegrini, M. & Sorbini, L. 1980b. Neotettonica di parte dei fogli Peschiera del Garda (48), Verona (49), Mantova (62) e di tutto il foglio Legnago (63). In CNR, Contributi alla realizzazione della Carta Neotettonica d'Italia. *P.F.Geodinamica*, 356: 613-655.

Baraldi, F., Magnani, T. & Zucchi, C. 1980c. *Inquinamento dei laghi di Mantova*. Mantova: Progetto Ambiente, Assessorato Sanità e Ambiente, Prov. Mantova.

Baraldi, F., Ferrario, G., Lugani, V. & Ottenziali, L. 1987. Recupero di aree degradate da attività estrattive nel terrazzo würmiano di Pozzolo (MN). *Proceedings of the VI Congr. O.N.G.*, Padova: 139-147.

Belenghi, C. 1908. *La ghiaia delle nostre cave (Collefiorito in Rivalta Mantovana)*. Mantova: Barbieri.

Bertazzolo, G. 1609. *Discorso sopra il nuovo sostegno di Governolo*. Mantova: Bibl. Comune di Mantova.

Carton, A., Cavallin, A., Francavilla, F., Mantovani, F., Panizza, M., Pellegrini, G.B. & Tellini, C. 1994. Ricerche ambientali per l'individuazione dei beni geomorfologici. Metodi ed esempi. *Il Quaternario*, 7(2): 365-372.

Castaldini, D. 1994. Pozzolo nell'Alta Pianura Mantovana. In Gruppo Nazionale Geografia Fisica e Geomorfologia. Proposta di legenda geomorfologica ad indirizzo applicativo. *Geogr. Fis. Din. Quat.*, 16: 143-144.

Castiglioni, G.B., Biancotti, A., Bondesan, M., Castaldini, D., Ciabatti, M., Cremaschi, M. & Favero, V. 1986. Criteri informativi del progetto di una carta geomorfologica della Pianura Padana. *Materiali*, 7: 1-29.

Castiglioni, G.B., Ajassa, R., Baroni, C., Biancotti, A., Bondesan, M., Brancucci, G., Castaldini, D., Cavallin, A., Cortemiglia, F., Cremaschi, M., Da Rold, O., Elmi, C., Fasani, D., Favero, V., Gasperi, G., Giorgi, G., Marchetti, G., Marchetti, M., Meneghel, M., Motta, M., Nesci, O., Orombelli, G., Paronuzzi, P., Pellegrini, G.B., Pellegrini, L., Tellini, C., Toniello, V., Turrini, M.C., Zecchi, R. & Zorzin, R. 1997a. *Carta altimetrica e dei movimenti verticali del suolo della Pianura Padana a scala 1:250,000*. Firenze: Selca.

Castiglioni, G.B., Ajassa, R., Baroni, C., Biancotti, A., Bondesan, A., Bondesan, M., Brancucci, G., Castaldini, D., Castellaccio, E., Cavallin, A., Cortemiglia, F., Cortemiglia, G.C., Cremaschi, M., Da Rold, O., Elmi, C., Favero, V., Ferri, R., Gandini, F., Gasperi, G., Giorgi, G., Marchetti, G., Marchetti, M., Marocco, R., Meneghel, M., Motta, M., Nesci, O., Orombelli, G., Paronuzzi, P., Pellegrini, G.B., Pellegrini, L., Rigoni, A., Sommaruga, M., Sorbini, L., Tellini, C., Turrini, M.C., Vaia, F., Vercesi, P.L., Zecchi, R. & Zorzin, R. 1997b. *Carta Geomorfologica della Pianura Padana a scala 1:250,000*. Firenze: Selca.

Cozzaglio, A. 1933. *Note illustrative della Carta Geologica d'Italia, Fogli 48 Peschiera e 62 Mantova*. Venezia: Uff. R. Magistr. Acque.

Cremaschi, M. 1987. *Paleosols and vetusols in the central Po Plain (Northern Italy)*. Milano: Unicopli.

Dal Ri, R. 1991. *La pianificazione delle attività di cava*. Roma: Autonomie.

Guzzetti, F., Marchetti, M. & Reichenbach, P. 1996. Large alluvial fans in the north-central Po Plain (Northern Italy). *Geomorphology*, 18: 119-136.

Istituto Geografico Militare Italiano (IGMI) 1885-1973. Carta Topografica d'Italia alla scala 1:25,000 e 1:100,000 del F. 62 Mantova, maps. Firenze: IGMI.

Marchetti, M., Panizza, M., Soldati, M. & Barani, D. (eds) 1995. Geomorphology and Environmental Impact Assessment. Proc. 1st and 2nd workshops of a "Human Capital and Mobility" project. *Quad. Geodinamica Alpina e Quaternaria*, 3: 1-197.

Panizza, M. 1996. *Environmental Geomorphology*. Amsterdam: Elsevier.

Panizza, M., Castaldini, D., Cremaschi, M., Gasperi, G. & Pellegrini, M. 1988. Ricostruzione paleogeografica e geodinamica tardopleistocenica ed olocenica dell'area centropadana tra Verona e Modena. In Ente Nazionale Energia Elettrica, *Contributi di preminente interesse scientifico agli studi di localizzazione di impianti nucleari in Piemonte e Lombardia*, 2: 1-410.

Panizza, M., Fabbri, A.G., Marchetti, M. & Patrono, A. (eds) 1996a. Special EU Project Issue: Geomorphology and Environmental Impact Assessment. *ITC Journ.*, 1995(4): 305-371.

Panizza, M., Marchetti, M. & Patrono, A. 1996b. A proposal for a simplified method for assessing impacts on landforms. In M. Panizza, A.G. Fabbri, M. Marchetti & A. Patrono (eds), Special EU Project Issue: Geomorphology and Environmental Impact Assessment. *ITC Journ.*, 1995(4): 324.

Petrucci, F. & Tagliavini, S. 1969. *Note illustrative della Carta Geologica d'Italia, Foglio 61, Cremona*. Roma: Geological Survey of Italy.

Pieri, M. & Groppi, G. 1981. Subsurface geological structure of the Po Plain, Italy. In CNR, Contributo alla realizzazione della Carta Neotettonica d'Italia, *Progetto Finalizzato Geodinamica*, 414: 1-13. Roma: CNR.

Regione Lombardia 1983. Carta Tecnica Regionale a scala 1:25,000 e 1:50,000, Regional Technical Map (C.T.R.), Milano: Regione Lombardia.

Slejko, D., Carraro, F., Carulli, G.B., Castaldini, D., Cavallin, A., Doglioni, C., Iliceto, V., Nicolich, R., Rebez, A., Semenza, E., Zanferrari, A. & Zanolla, C. 1987. *Modello sismotettonico dell'Italia nord-orientale*. Trieste: Ricci.

Veggiani, A. 1974. Le variazioni idrografiche del basso corso del Po negli ultimi 3000 anni. *Padusa*, 1-2: 23-45.

Venzo, S. 1965. Rilevamento geologico dell'anfiteatro morenico frontale del Garda, dal Chiese all'Adige. *Mem. Soc. It. Sci. Nat.*, 14(1): 1-81.

Zanferrari, A., Bollettinari, G., Carobene, L., Carton, A., Carulli, G.B., Castaldini, D., Cavallin, A., Panizza, M., Pellegrini, G.B., Pianetti, F. & Sauro, U. 1982. Evoluzione neotettonica dell'Italia Nord-Orientale. *Mem.Sc.Geol.*, 35: 355-376.

Prediction models for landslide hazard zonation using a fuzzy set approach

C.F. Chung
Geological Survey of Canada, Canada

A.G. Fabbri
International Institute of Aerospace Survey and Earth Sciences (ITC), The Netherlands

ABSTRACT: Quantitative models for hazard zonation are based on a spatial data base consisting of several layers of digital maps which are integrated using a geographical information system (GIS). A quantitative prediction model has been proposed for landslide hazard zonation using Fuzzy set theory. In this contribution, five estimation procedures using different fuzzy set operators in addition to a simple "direct" procedure, are compared and evaluated. The model assumes that future landslides can be predicted by the quantitative relationships established between past landslides, representing a mass movement of a given dinamic type, and the spatial data base - the layers of maps which are considered to be representative of the conditions of landslide occurrences.

The prediction of landslide hazard at a given point (or sub-area) is considered as a fuzzy set that the point is likely to be affected by a future landslide given the spatial data at the point. The fuzzy set operators for the procedures which will be compared in terms of prediction power are: a) minimum (intersection); b) maximum (union); c) algebraic product; d) algebraic sum; and e) gamma operator.

A case study from an area in central Italy, which is affected by earth and debris flows is used to compare the fuzzy set procedures in addition to the direct one. In the spatial data base, it was pretended that the time of the study was the year 1955 and that all the spatial data available in 1955 were compiled including the distribution of the landslides which had occurred prior to that year. These were used as the training data set to construct the membership functions between the landslides and the remainder of the input data set. The predictions based on those membership functions were then evaluated by comparing the map pattern of the predicted hazard classes with the distribution of the landslides which had occurred after 1955, i.e. during the period 1956-1993.

1 INTRODUCTION

The quantitative fuzzy set model presented here for the prediction of landslide hazard is based on the "favourability function" proposed by Chung & Fabbri (1993) which requires the following two assumptions: a) that past landslides of a given type (i.e. a clearly identified type of process) can be characterized by sets of layers of supporting spatial data, and b) that landslides of the same type will occur in the future under similar circumstances. The landslide hazard prediction at each point in the data set is considered as a member of the fuzzy set consisting of the fact that the point will be affected by a future landslide given the information from the supporting spatial input data at the point.

Bayesian formulas for geologic prediction models were used by Spiegelhalter (1986) and Agterberg et al. (1990). Chung & Fabbri (1993) and Fabbri & Chung (1996) have adapted the formulas to geomorphologic hazard zonation as a part of "favourability function" approaches in predictive spatial data analysis. The method has been applied to landslide prediction by Chung & Leclerc (1994), Leclerc (1994), Leclerc & Chung (1993), Luzi (1995), and Luzi & Fabbri (1995). Multivariate regression analysis for landslide hazard was proposed earlier by Carrara (1983), Carrara et al. (1992), and Chung et al. (1995). Generally, in the regression model, a square grid is first overlaid on a study area, and its size is determined so that the total number of grid cells (or pixels) is reasonably small (generally only a few thousands). A prerequisite for this traditional cell-based regression analysis is a rasterization of the spatial input. Chung et al. (1995) have extended the cell-based regression to a regression model based on "unique-condition" sub-areas. This model, which used weighted regression did not require rasterization and for that reason provided greater generality of data structure. The use of regression analysis without imposing rasterization is an important and significant step when the data is in a vector-based GIS. Chung & Fabbri (1998) have also presented a probabilistic prediction model for landslide hazard mapping. They have proposed five estimation procedures for that prediction model and compared their results to identify an optimal procedure.

In applying the fuzzy set model to landslide hazard zonation, several input maps are used and overlaid with the map of all the landslides known to have occurred before a given time, i.e. that have been recognized and mapped. The map of past landslides, or past events, is normally in a binary form, just indicating the presence of the occurrences. The final product of modeling is a hazard map. If more recent landslides of the same type have occurred, the distribution of the different values in the hazard map can be compared with their distribution. In this manner, comparative analysis is performed in which different models and several modifications of the input data can be evaluated in a systematic way. In addition, sensitivity analysis can also be done using a similar technique.

2 STUDY AREA AND DATA

The input data for quantitative landslide hazard zonation usually consist of several layers of primary or compiled maps. Each layer may be the result of map updating by experts, of field verification and of interpretation of aerial photographs. The maps describe surficial and bedrock engineering geology, soil type, slope, land-use, aspect, drainage patterns, and mass movements, which are known to be relevant to slope instability. In addition, the identification of types (corresponding to specific geomorphologic settings) and dates of landslide phenomena (allowing separation of periods of activity) are critical to the application of predictive techniques.

Among many layers of spatial data constructed by Luzi (1995), in the Fabriano study area in central Italy, used also in this contribution (geographical coordinates of the area are in the illustrations), she has suggested that the six data layers: a) engineering geology map; b) slope angle map; c) aspect map; d) land-use map, and two maps expressing the distance from the nearest drainage lines e), and from faults f), are significantly related to landslide hazard. The corresponding classes in each layer are shown in the first column of Table 2. In the spatial data base it was pretended that the time of the study was the year 1955. All the seven layers available from 1955 were compiled in addition to the distribution of the landslides which had occurred prior to that year. These were used as the training data set to construct probabilistic relationships between the landslides and the remainder of the input data set, i.e. the six map layers. The predictions based on those relationships were then evaluated by comparing the map pattern of the predicted hazard classes with the distribution of the landslides which had occurred after 1955, i.e. during the period 1956-1993. The landslides were all mapped in the field and verified by photo-interpretation using aerial photographs taken in 1955 and in 1992, respectively. We have used the six layers in Luzi (1995) to predict and map landslide hazard.

3 DATA FORMAT, QUANTIZATION AND UNIQUE-CONDITION SUB-AREAS

Although some of the six layers used in this application represent continuous measurements such as slope angles and distances, as discussed by Chung et al. (1995), a map layer containing continuous measurements is usually converted into a number of classes, "thematic classification", for producing a new map representing geologic hazard. As in the Fabriano study area, we may assume that each layer represents a classification map containing a number of thematic classes.

For convenience in this paper, to capture map information in digital form, we use a grid of square-cells or "raster" data structure rather than a "vector" data structure (Aronoff 1989, Chung & Fabbri 1993). In the raster structure, a square grid is overlaid on each map. Each grid cell is termed a "pixel" (picture element) and each map is represented by a rectangular matrix of numbers, each representing a pixel in one-to-one correspondence with a small area on the map. In the Fabriano study area, each map, obtained by digitizing 1:50,000 and 1:25,000 maps, consisted of 922 pixels x 515 pixels matrix of numbers (474,830 pixels), and each pixel corresponded to a 30 m x 30 m area on the ground. For instance, the value of the pixels for the engineering geology map, represented the class of bedrock lithology in the corresponding small area on the ground.

Let us denote as **A** the whole study area and let us consider the kth layer, L_k containing thematic classification data such as landslide-related bedrock lithology. Without loss of generality, we may assume that the quantized value at a pixel p in L_k takes one integer value among $\{1, 2, \cdots, n_k\}$ where n_k is the maximum number of the classes in L_k (i.e. in L_k, **A** is divided into n_k non-overlapping sub-areas). One of these n_k units may represent the areas not covered by the observed values in the layer (i.e. it may represents an "unmapped" class).

Then, the m layers of map data at every pixel p in **A** are represented by:

$$v_k(p) = \{(v_1(p), \cdots, v_m(p)), p \in \mathbf{A}\} \quad (3.1)$$

where $v_k(p)$ is the quantized value for the kth layer L_k at p.

Here the m quantized values $(v_1(p), \cdots, v_m(p))$ in (equation 3.1) are regarded as m pieces of evidence toward establishing whether p is likely to be affected by a future landslide, assuming what stated earlier in a) and b). We may regard the quantization, v_k, as a function of **A** into a finite interval for the kth layer:

$$v_k : \mathbf{A} \rightarrow \{1, 2, \cdots, n_k\} \quad (3.2)$$

Let us consider the sub-area covered by all pixels whose values $(v_1(p), \cdots, v_m(p))$ are identical. Such a set of all pixels where the observed values of the m input layers are identical is termed the "unique-condition sub-area" (the term "unique-condition" polygon is generally used in vector-based GIS data structures, but it is applicable to raster data structures as well). The whole study area can be subdivided into a number (much smaller than the total number of pixels) of the non-overlapping unique condition sub-areas. For the Fabriano study area, we have 58,496 unique-condition sub-areas compared to 474,830 pixels.

The distribution map in Figure 1 shows the translational landslides of combined type "earth flows and debris flows" (Luzi 1995), or EDF in short, which occurred prior to the year 1955. A data base constructed by Luzi (1995) separates in time those landslides from the later ones. Figure 1 also illustrates the distribution of the EDF landslides, translational mass movements which occurred during 1956-1993 in the same study area. Only the pre-1955 landslides, were used for constructing the probability prediction models. The 1956-1993 EDF landslides were used for cross-validation of the prediction models estimated by the pre-1955 occurrences.

4 FUZZY SET MODEL

Consider a pixel p in **A** and the following proposition:

T_p : "p *will be affected by a future landslide of type* EDF" (4.1)

At a pixel p in **A**, we have the information on the m layers, $(v_k(p), k=1, \cdots, m)$. The problem is to define a function at every pixel p representing a degree of support for the proposition that p is likely to be affected by a future EDF landslide given the m evidences $(v_k(p), k=1, \cdots, m)$ at p, i.e. to define

$f(T_p|\text{Given m evidences } v_k(p), k=1, \cdots, m)$ at every p in **A** (4.2)

Chung & Fabbri (1993) have defined such a function as the "favourability function" for the proposition T_p discussed in expression (4.1).

In this contribution, we will discuss only one type of the favourability function in (4.2) which is the membership function of Zadeh's fuzzy set (Zadeh 1965, 1968) consisting of all pixels which are likely to be affected by a future landslide of type EDF given the m evidences at a pixel i.e.

$$f(T_p|v_1(p), v_2(p), \cdots, v_m(p)) = \mu\{p|v_1(p), v_2(p), \cdots, v_m(p)\} \quad (4.3)$$

where $\mu\{p|v_1(p), v_2(p), \cdots, v_m(p)\}$ is the membership function of the fuzzy set.

4.1 Basic idea of Zadeh's Fuzzy set model

Fuzzy set **S** consists of all the pixels ($p \in $ **A** : whole area) where the proposition in expression (4.1) is "likely" to be true. **S** is defined by a membership function μ_s:

$$\mu_s : \mathbf{A} \to [0, 1] \tag{4.4}$$

It represents a "*grade of membership*", or a "*degree of compatibility*". We may also interpret the membership function as a "possibility function" of p where the proposition is possibly true. Conversely, the fuzzy set **S** is also uniquely defined by the membership function μ_s:

$$\mathbf{S} = \{(p, \mu_s(p)), p \in \mathbf{A}\} \tag{4.5}$$

The definitions in (4.4) and (4.5) imply that the fuzzy set and the corresponding membership function are equivalent. Usually a fuzzy set is defined by a conceptual idea, and then the corresponding membership function is constructed to determine a "degree of support" for each member. In practice, the membership function (also "*possibility function*", Zadeh 1978), μ_s is difficult to construct. Our prediction problem can be represented in the following task: "construct the membership function at each pixel in expression (4.4) after observing the m evidences, ($v_k(p)$, k=1, $\cdots$, m) at p in **A**, as a function describing the support for the condition that p is likely to be affected by a future EDF landslide".

4.2 Construction of membership function

Consider a pixel p with the m evidences, $v_1(p) = c_1, \cdots, v_m(p) = c_m$. To construct the membership function $\mu\{p|v_1(p) = c_1, \cdots, v_m(p) = c_m\}$ at pixel p, we use the following terms. Let **A** denote the whole study area. Let "*mes* of (B)" represent the surface area covered by any sub-area B in **A**. Suppose that **P** denotes the areas affected by the past landslides in **A** and **F** denotes the unknown areas which will be affected by landslides (**F** for future and **P** for past events).

Consider the kth layer. Suppose that we have n_k classes (map units), $\{1, 2, \cdots, n_k\}$. It means that the whole area **A** is divided into n_k non-overlapping sub-areas using the kth layer, L_k. Let A_{kr} denote the sub-area covered by all the pixels whose class is equal to r in the kth layer, L_k where r takes one of $\{1, 2, \cdots, n_k\}$ values. Hence **A** consists of non-overlapping n_k sub-areas $\{A_{k1}, A_{k2}, \cdots, A_{kn_k}\}$ in the kth layer. Let $\mathbf{F} \cap A_{kr}$ be the unknown area to be affected by future landslides within A_{kr} and $\mathbf{P} \cap A_{kr}$ represent the area affected by the past landslides within A_{kr}.

Hence $\bigcap_{k=1}^{m} A_{kc_k}$ is the unique condition sub-area where the m pixel values of the m layers at each pixel are $(c_1, \cdots, c_m)$, and $\bigcap_{k=1}^{m} (\mathbf{F} \cap A_{kc_k})$ denotes the unknown area which is to be affected by future landslides within the unique-condition sub-area, $\bigcap_{k=1}^{m} A_{kc_k}$. Then one way of constructing the membership function can be:

$$\mu\{p|v_1(p) = c_1, \cdots, v_m(p) = c_m\} = \textit{mes of } \{\bigcap_{k=1}^{m} (\mathbf{F} \cap A_{kc_k})\} / \textit{mes of } (\bigcap_{k=1}^{m} A_{kc_k})$$

$$\mathbf{S} = \{(p, \mu\{p|v_1(p) = c_1, \cdots, v_m(p) = c_m\}), p \in \mathbf{A}\} \tag{4.6}$$

We present the five procedures to estimate the membership function in expression (4.6). To evaluate the results obtained by the procedures, it was established that the time of the study was the year 1955 and that all the spatial data available from 1955 were compiled including the distribution of the landslides in Figure 1, which occurred prior to 1955. The pre-1955 landslide data were used as the training data set to construct the membership relationships between the past landslides and the input data set of supporting digital maps. The prediction based on these relationships is then evaluated by analyzing the distribution, shown in Figure 1, of the landslides which occurred during the period 1956-1993. (fig. 1, see color plate pages, following p. 225)

4.3 Direct procedure and visualization

Before we present the five fuzzy set procedures and the associated fuzzy set operators, we first must discuss a simple procedure to construct the membership function in expression (4.6). This estimated membership function will be used as a yard stick to evaluate the prediction procedures presented in the following sections.

Consider a pixel p in **A** with $v_k(p) = c_k$ for $k=1, \cdots, m$. It means that the thematic class of the k-th layer of the pixel p is c_k for $k=1, \cdots, m$. The simplest estimate of the membership function in (4.6) can be obtained by assuming that:

$$\bigcap_{k=1}^{m} (\mathbf{F} \cap A_{kc_k}) = \bigcap_{k=1}^{m} (\mathbf{P} \cap A_{kc_k}) \quad (4.7)$$

where $\bigcap_{k=1}^{m} (\mathbf{P} \cap A_{kc_k})$ denotes the area affected by the past landslides within the unique condition sub-area, $\bigcap_{k=1}^{m} A_{kc_k}$, where every pixel has the m pixel values, $(c_1, \cdots, c_m)$ for the m layers. The first estimator of the membership function is (the subscript d indicates the EDF landslides):

$$\hat{\mu}_d \{p|v_1(p) = c_1, \cdots, v_m(p) = c_m\} = mes \text{ of } \{\bigcap_{k=1}^{m} (\mathbf{P} \cap A_{kc_k})\} / mes \text{ of } (\bigcap_{k-1}^{m} A_{kc_k}) \quad (4.8)$$

For the Fabriano data, **P** is the set of pixels affected by the pre-1955 EDF landslide occurrences shown in black in Figure 1, and these pixels are used to compute the membership function by expression (4.8). The set **P** is used as "a training data-set" to establish the membership relationships between the past landslides and the input spatial data set.

The prediction patterns based on the membership function $\hat{\mu}_d = \{p|c_1, \cdots, c_m\}$ in expression (4.8) are shown in Figure 2. Those prediction patterns are compared with the pre-1955 EDF landslide occurrences shown in black in Figure 1 to obtain the success rates illustrated in both Table 1a and the corresponding graph (brown line) in Figure 5a. The prediction patterns are also compared with the 1956-1993 EDF landslide occurrences shown in red in Figure 1 to obtain the prediction rates illustrated in both Table 1b and in the corresponding graph (brown line) in Figure 5b.(fig. 2, see color plate pages, following p. 225)

Table 1a. Success rates for pre-1955 EDF landslide occurrences (training data). The membership functions and fuzzy set models were based on pre-1955 data, Fabriano study area, central Italy. Corresponding graphs are shown in Figure 5a.

Row number	Area (%)	Class	Color	Direct unmodif. (I)	Fuzzy: minimum operator (II)	Fuzzy: maximum operator (III)	Fuzzy: algebraic product $\gamma = 0$ (IV)	Fuzzy: algebraic sum $\gamma = 1$ (V)	Fuzzy: gamma $\gamma = 0.5$ (VI)
1	0.05	0-5	purple red	0.725	0.279	0.219	0.019	0.362	0.000
2	0.10	0-10	red	0.931	0.471	0.424	0.096	0.593	0.111
3	0.15	0-15	orange	0.998	0.626	0.665	0.205	0.764	0.261
4	0.20	0-20	yellow	1.000	0.826	0.874	0.343	0.880	0.409
5	0.25	0-25	light green	1.000	0.927	0.890	0.478	0.919	0.512
6	0.30	0-30	green	1.000	0.964	0.904	0.591	0.935	0.615
7	0.35	0-35	blue-green	1.000	0.972	0.920	0.715	0.948	0.706
8	0.40	0-40	cyan	1.000	0.990	0.929	0.808	0.958	0.782
9	0.45	0-45	light blue	1.000	0.996	0.939	0.879	0.966	0.848
10	0.50	0-50	blue	1.000	0.997	0.947	0.933	0.975	0.899

For visualization purposes in Figure 2, the pixels were divided into four classes according to the values computed by $\hat{\mu}_d \{p|c_1, \cdots, c_m\}$ in expression (4.8). Sub-areas of 5% of the study area were arbitrarily selected as unit thresholds of the predicted values which range between 0 and 1. The pixels shown in "purple red" in the illustration, with the highest 5% estimated values among all the pixels in the area were classified as the class "0-5%"

corresponding to Column I ("Direct") of Row 1 ("0-5%") in Tables 1a and 1b. They occupy 5% of the whole study area **A**. The pixels shown in "red" in Figure 2, with the next highest 5% values were added to the "0-5%" class and they are together classified as the class "0-10%" corresponding to Column I / Row 2 in Tables 1a and 1b. We repeated the classification once more to obtain class "0-15%". The fourth class in Figure 2 contains the remaining values, class ">15%". The first three classes obtained correspond to the first three rows in Column I in Tables 1a and 1b.

While Column I ("Direct") in Table 1a shows the success rates of the pre-1955 occurrences, Column I ("Direct") in Table 1b shows the prediction rates for the 1956-1993 occurrences. These two columns are represented as brown lines in Figures 5a and 5b. From Figure 5a, obviously the success rates of the membership function in expression (4.8) has the best performance among all the membership functions considered. The prediction rate performed badly, however, as shown in Figure 5b.

Table 1b. Prediction rates for 1956-1993 EDF landslide occurrences. The Membership functions and fuzzy set models were based on pre-1955 data, Fabriano study area, central Italy. Corresponding graphs are shown in Figure 5b.

Row number	Area	Class (%)	Color	Direct unmodif. (I)	Fuzzy: minimum operator (II)	Fuzzy: maximum operator (III)	Fuzzy: algebraic product $\gamma = 0$ (IV)	Fuzzy: algebraic sum $\gamma = 1$ (V)	Fuzzy: gamma, $\gamma = 0.5$ (VI)
1	0.05	0-5	purple red	0.165	0.151	0.134	0.003	0.174	0.018
2	0.10	0-10	red	0.377	0.354	0.404	0.030	0.410	0.049
3	0.15	0-15	orange	0.483	0.554	0.648	0.066	0.632	0.105
4	0.20	0-20	yellow	0.508	0.676	0.826	0.157	0.821	0.216
5	0.25	0-25	light green	0.536	0.836	0.847	0.261	0.863	0.325
6	0.30	0-30	green	0.568	0.921	0.862	0.405	0.897	0.448
7	0.35	0-35	blue-green	0.605	0.921	0.886	0.552	0.916	0.563
8	0.40	0-40	cyan	0.652	0.951	0.895	0.681	0.936	0.667
9	0.45	0-45	light blue	0.699	0.955	0.900	0.785	0.952	0.750
10	0.50	0-50	blue	0.730	0.965	0.910	0.883	0.960	0.824

Column I / Row 1 in Tables 1a and 1b contains two values 0.725 and 0.165, respectively. This means that the 0-5% class, with the highest shown as "purple red" in Figure 2 contains 72.5% of the pre-1955 occurrences and 16.5% of the 1956-1993 occurrences, while this class occupies 5% of the whole study area **A**. If we select 5% of the whole area randomly, then we would expect that the randomly selected area contains about 5% of the pre-1955 occurrences and also about 5% of the 1956-1993 occurrences (should they be randomly distributed). Obviously 72.5% and 16.5% are much higher than the expected 5%. The value of 72.5% illustrates that the prediction patterns in Figure 2 are a good representation of the pre-1955 training data set in terms of the input spatial data set, and this means that the establishment of quantitative relationships between the landslides (pre-1955) which have occurred and the input spatial data set was successful. This class predicted only 16.5%, however, of the landslide occurrences to come in the next 38 years (1956-1993).

As a prediction model for future landslides, the prediction rates with respect to the 1956-1993 EDF landslides are the only significant statistics. Consider Column I / Row 3 in Tables 1a and 1b containing the values 0.998 and 0.483, respectively. The value 48.3% is the prediction rate for the "future" (with respect to year 1955) 1956-1993 EDF landslides. The set of the pixels in these three classes (purple red, red, and orange colored pixels) occupies 15% of the whole area, but it predicted 48.3% of the EDF landslides to come in the next 38 years (1956-1993). As a prediction tool, Figure 2 produced a reasonable result. The relatively poor performance (48.3% vs. 99.8%) for future EDF landslides may imply that this direct procedure for the fuzzy membership function in (4.6) should be improved.

5 THE FIVE FUZZY SET OPERATORS

Consider a pixel p with the m evidences, $v_1(p) = c_1, \cdots, v_m(p) = c_m$. To estimate the membership function $\mu_S \{p|v_1(p) = c_1, v_2(p) = c_2, \cdots, v_m(p) = c_m\}$ at pixel p in (4.6), we first define m fuzzy sets and the m corresponding membership functions, one for each layer. At each kth layer $\mathbf{L}_k$, we define a fuzzy set $\mathbf{S}_k$ and the corresponding membership function $\mu_{S_k}(p)$ based on the pixel value $v_m(p) = c_m$ in $\mathbf{L}_k$:

$$\mu_{S_k} : \mathbf{A} \to [0, 1] \qquad \mathbf{S}_k = \{(p, \mu_{S_k}\{p|v_k(p)\}), p \in \mathbf{A}\} \qquad (5.1)$$

The favourability function $f_k(p) = \mu_{S_k}\{p|v_k(p)\}$ for the kth layer $\mathbf{L}_k$ represents the degree of support that the pixel p belongs to a fuzzy set consisting of pixels which are likely to be affected by a future EDF landslide, given the only one evidence, $v_k(p) = c_k$ in the kth layer. We repeat it m times, one for each layer and obtain the following m fuzzy sets and the corresponding m membership functions:

$$\mathbf{S}_1 = \{(p, \mu_{S_1}\{p|v_1(p)\}), p \in \mathbf{A}\}$$
$$\mathbf{S}_2 = \{(p, \mu_{S_2}\{p|v_2(p)\}), p \in \mathbf{A}\}$$
$$\cdots \qquad (5.2)$$
$$\mathbf{S}_m = \{(p, \mu_{S_m}\{p|v_m(p)\}), p \in \mathbf{A}\}$$

It is much easier to construct the membership function $\mu_{S_k}\{p|v_k(p)\}$ than the membership function $\mu_S\{p|v_1(p), v_2(p), \cdots, v_m(p)\}$, because only one evidence, $v_k(p) = c_k$ at p in the kth layer need to be considered.

As in (4.6), one simple but reasonable way to obtain the kth membership function is:

$$\mu_{S_k}\{p|v_k(p) = c_k\} = mes \text{ of } (\mathbf{F} \cap \mathbf{A}_{kc_k}) \,/\, mes \text{ of } \mathbf{A}_{kc_k} \qquad (5.3)$$

where $\mathbf{A}_{kc_k}$ is the area covered by c_k-thematic class, and $\mathbf{F} \cap \mathbf{A}_{kc_k}$ is the unknown area affected by future landslides within $\mathbf{A}_{kc_k}$ in the kth layer.

As in (4.7), the simplest estimate of the membership function in (5.3) can be obtained by assuming that :

$$\mathbf{F} \cap \mathbf{A}_{kc_k} = \mathbf{P} \cap \mathbf{A}_{kc_k} \qquad (5.4)$$

where $\mathbf{P} \cap \mathbf{A}_{kc_k}$ denotes the area affected by the past EDF landslides within $\mathbf{A}_{kc_k}$ where every pixel has the same pixel value, c_k in the kth layer. Then the kth membership function in (5.3) and the corresponding fuzzy set are estimated by:

$$\hat{\mu}_{S_k}\{p|v_k(p) = c_k\} = mes \text{ of } (\mathbf{P} \cap \mathbf{A}_{kc_k}) \,/\, mes \text{ of } \mathbf{A}_{kc_k}$$
$$\hat{\mathbf{S}}_k = \{(p, \hat{\mu}_{S_k}\{p|v_k(p)\}), p \in \mathbf{A}\} \qquad (5.5)$$

Another way to obtain the kth membership function in (5.3) is by using or introducing expert's knowledge directly and we denote the expert's estimator of the kth membership function and the corresponding fuzzy set as:

$\tilde{\mu}_{S_k}\{p|v_k(p) = c_k\}$ is determined by the expert's knowledge,

$$\tilde{\mathbf{S}}_k = \{(p, \tilde{\mu}_{S_k}\{p|v_k(p)\}), p \in \mathbf{A}\} \qquad (5.6)$$

We will construct the membership function $\mu_S\{p|v_1(p), v_2(p), \cdots, v_m(p)\}$ in (4.6) by combining the m membership functions based on the past landslides in (5.5) or the m membership functions based on the expert's knowledge in (5.6) using the fuzzy set operators. Let us first discuss the five basic fuzzy set operators based on two fuzzy sets, $\mathbf{S}_1$ and $\mathbf{S}_2$ and the corresponding two membership functions, μ_1 and μ_2. Then:

a) Complement operator: $S = S_1^c \Leftrightarrow \mu_S = 1 - \mu_{S_1}$

b) Intersection (minimum) operator: $S = S_1 \cap S_2 \Leftrightarrow \mu_S = \min(\mu_{S_1}, \mu_{S_2})$

c) Union (maximum) operator: $S = S_1 \cup S_2 \Leftrightarrow \mu_S = \max(\mu_{S_1}, \mu_{S_2})$

d) Algebraic product operator: $S = S_1 \otimes S_2 \Leftrightarrow \mu_S = \mu_{S_1} \cdot \mu_{S_2}$

e) Algebraic sum operator: $S = S_1 \oplus S_2 = (S_1^c \otimes S_2^c)^c \Leftrightarrow \mu_S = \mu_{S_1} + \mu_{S_2} - \mu_{S_1} \cdot \mu_{S_2}$

The latter four operators can be generalized for m fuzzy sets, $(S_1, S_2, \cdots, S_m)$ and the corresponding m membership functions, $(\mu_{S_1}, \mu_{S_2}, \cdots, \mu_{S_m})$, are shown in (5.7), (5.8), (5.9) and (5.10), respectively. In addition, we consider another operator in (5.11) which is a function of two fuzzy set operators, the algebraic product and the algebraic sum operators and that has a parameter, γ. From (5.11), although it is simple and clear how to obtain the membership function, the exact meaning of the corresponding fuzzy set S is unknown.

1) Intersection (minimum) operator: $S = \bigcap_{k=1}^{m} S_k \Leftrightarrow \mu_S = \min(\mu_{S_1}, \mu_{S_2}, \cdots, \mu_{S_m})$ (5.7)

2) Union (maximum) operator: $S = \bigcup_{k=1}^{m} S_k \Leftrightarrow \mu_S = \max(\mu_{S_1}, \mu_{S_2}, \cdots, \mu_{S_m})$ (5.8)

3) Algebraic product operator: $S = \bigotimes_{k=1}^{m} S_k \Leftrightarrow \mu_S = \prod_{k=1}^{m} \mu_{S_k}$ (5.9)

4) Algebraic sum operator:

$$S = \bigoplus_{k=1}^{m} S_k = \left(\bigotimes_{k=1}^{m} S_k^{\,c} \right)^c \Leftrightarrow \mu_S = \sum_{k=1}^{n} \mu_{S_k} - \sum_{k=1}^{n} \sum_{j=k+1}^{n} \mu_{S_k} \mu_{S_j} + \cdots + (-1)^m \prod_{k=1}^{m} \mu_{S_k} = 1 - \prod_{k=1}^{m} \left(1 - \mu_{S_k}\right) \quad (5.10)$$

5) Gamma operator:

$$S = f\left(\bigoplus_{k=1}^{m} S_k, \bigotimes_{k=1}^{m} S_k \middle| \gamma \right) \Leftrightarrow \mu_S = \left[\prod_{k=1}^{m} \mu_{S_k} \right]^{1-\gamma} \left[1 - \prod_{j=1}^{m} \left(1 - \mu_{S_j}\right) \right]^{\gamma} \quad (5.11)$$

where $0 \leq \gamma \leq 1$ and:
Gamma operator with $\gamma = 1$ is an algebraic sum operator;
Gamma operator with $\gamma = 0$ is an algebraic product operator.

When one defines a *Fuzzy event* instead of a *Fuzzy set* using a similar membership function, then the *probability function* has the relationship:

$$\Pr\{S\} = E(\mu_S) \quad (5.12)$$

where $E(\mu_S)$ is the expectation (Bickel & Doksum 1977). In addition, we have:

$$\Pr\{S_1 \cup S_2\} = \Pr\{S_1\} + \Pr\{S_2\} - \Pr\{S_1 \cap S_2\}$$

$$\Pr\{S_1 \oplus S_2\} = \Pr\{S_1\} + \Pr\{S_2\} - \Pr\{S_1 \cdot S_2\}$$

The five operators discussed in (5.7), (5.8), (5.9), (5.10), and (5.11) will be applied to the estimated membership functions defined in either (5.5) or (5.6).

6 FIVE PROCEDURES

As before, let us consider a pixel p in A with $v_k(p) = c_k$ for $k=1, \cdots, m$. The thematic class of the k-th layer of the pixel p is c_k for $k=1, \cdots, m$. For the following five sections, we are going to use, for every layer, the estimated k^{th} membership function, and the corresponding fuzzy set in (5.5) are used:

$$\hat{\mu}_{S_k}\{p|v_k(p) = c_k\} = \text{mes of } (\mathbf{P} \cap A_{kc_k}) / \text{mes of } A_{kc_k}$$

$$\hat{S}_k = \{(p, \hat{\mu}_{S_k}\{p|v_k(p) = c_k\}), p \in \mathbf{A}\} \quad (6.0)$$

6.1 *Intersection (minimum) operator*

We obtain the second estimator of the fuzzy set S introduced in expression (4.6) by using the minimum operator defined in (5.7) and the m membership functions and the corresponding fuzzy sets in (6.0):

$$\hat{S} = \bigcap_{k=1}^{m} \hat{S}_k \Leftrightarrow \hat{\mu}_S\{p|c_1,\cdots,c_m\} = \min(\hat{\mu}_{S_1}\{p|c_1\},\cdots,\hat{\mu}_{S_m}\{p|c_m\}) \tag{6.1}$$

The prediction patterns based on the membership function $\hat{\mu}_S\{p|c_1,\cdots,c_m\}$ in expression (6.1) are shown in Figure 3. The prediction patterns are compared with the pre-1955 EDF landslides occurrences shown in black in Figure 1 to obtain the success rates illustrated in both Table 1a and the corresponding graph (green line) in Figure 5a. The prediction patterns shown in Figure 3 are also compared with the 1956-1993 EDF landslides occurrences shown in red in Figure 1 to obtain the prediction rates illustrated in both Table 1b and the corresponding graph (green line) in Figure 5b.

In Table 1, the pixels were divided into 10 classes (purple red to yellow to blue in color) according to $\hat{\mu}_S\{p|c_1,\cdots,c_m\}$ and are shown in Figure 3. The pixels ("purple red" in Figure 3) with the highest 5% estimated values among all the pixels in the area were classified as the "0-5%" class, corresponding to Column II ("minimum") of Row 1 in Tables 1a and 1b. The pixels (red in Figure 3) with the next highest 5% values were added to the 0-5% class and classified as the "0-10%" class, corresponding to Column II / Row 2 in Tables 1a and 1b. We repeated the classification nine more times and the 11 classes are represented by the values in Column II in Tables 1a and 1b. (fig. 3, see color plate pages, following p. 225)

Column II in Table 1a shows the success rates for (6.1) with respect to the pre-1955 EDF occurrences. The success rates are also shown as a green line in Figure 5a. Column II in Table 1b shows the prediction rates for (6.1) with respect to the 1956-1993 EDF occurrences and the same prediction rates are also shown as a green line in Figure 5b. Column II / Row 2 in Tables 1a and 1b contain the two values 0.471 and 0.354. The 0-10% class contains 47.1% of the pre-1955 occurrences and 35.4% of the 1956-1993 occurrences, while it occupies only 10% of the whole area. The values 47.1% and 35.4% are much better than the expected 10%. Let us compare these with the results of the direct procedure using (4.8). Although the success rate of 47.1% for using the minimum operator is much worse than the 93.1% from the direct procedure in (4.8) for the pre-1955 EDF occurrences, the prediction rates of 35.4% and 37.7% from the two procedures are within the same range and are much better than the 10% expected prediction rate.

Column II / Row 5 ("0-25%") in Tables 1a and 1b show two percentiles, 92.7% and 83.6%, respectively. All the pixels in this cumulative fifth class occupy 25% of the area, but they contain 92.7% of the pre-1955 EDF landslide occurrences and correctly predict 83.6% of the EDF landslides to come for the next 38 years (1956-1993). In this class, the prediction rate (83.6%) from the minimum operator is much greater than the prediction rate (53.6%) from the direct procedure for the 1956-1993 EDF landslides.

Obviously, the direct procedure in (4.8) produced much better results than the minimum operator in (6.1) for the pre-1955 training data set (success rates). The prediction rates for the 1956-1993 EDF landslide occurrences, however, except for the first two classes, show that the minimum operator produced better results than the ones obtained from (4.8).

6.2 *Union (maximum) operator*

We obtain the third estimator of the fuzzy set S discussed in (4.6) by using the maximum operator defined in (5.8) and the m membership functions and the corresponding fuzzy sets in (6.0):

$$\hat{S} = \bigcup_{k=1}^{m} \hat{S}_k \Leftrightarrow \hat{\mu}_S\{p|c_1,\cdots,c_m\} = \max(\hat{\mu}_{S_1}\{p|c_1\},\cdots,\hat{\mu}_{S_m}\{p|c_m\}) \tag{6.2}$$

For the prediction patterns based on the estimated membership function $\hat{\mu}_S\{p|c_1,\cdots,c_m\}$ in (6.1), the pixels were divided into 11 classes (purple red to yellow to blue in color) according to $\hat{\mu}_S\{p|c_1,\cdots,c_m\}$ similar to the minimum operator in the previous section. The pixels with the highest 5% estimated values among all the pixels in the area were classified as "0-5%" corresponding to Column III ("Maximum") of Row 1 in Tables 1a and 1b. The pixels with the next highest 5% values were added to the 0-5% class and classified as the "0-10%" class corresponding to Column III / Row 2 in Tables 1a and 1b. We repeated the classification nine times and the 11 classes are corresponding to the values in Column II in Tables 1a and 1b.

Column III in Tables 1a and 1b show the *success rates* and the *prediction rates* of expression (6.2) with respect to the pre-1955 occurrences and 1956-1993 occurrences.

These two rates are also represented in Figures 5a and 5b, respectively, as blue lines. Column III / Row 3 in Tables 1a and 1b contain two values 0.665 and 0.648. The 0-15% class contains 66.5% of the pre-1955 occurrences and 64.8% of the 1956-1993 occurrences, while it occupies 15% of the whole area. The prediction rate of 64.8% is better than 55.5% from the minimum operator discussed in the previous section and is much better than the expected 15%.

Column III / Row 4 ("0-20%") in Tables 1a and 1b show two percentiles 0.874 and 0.826. All the pixels in this cumulative fourth class occupy 20% of the area, but contain 87.4% of the pre-1955 occurrences and correctly predicted 82.6% of the landslides to come in the next 38 years (1956-1993). Comparing Column III in Tables 1a and 1b, the prediction rates for the 1956-1993 data are about the same as the success rates for the training data of the pre-1955 EDF landslides.

6.3 *Algebraic product operator*

We obtain the third estimator of the fuzzy set S introduced in expression (4.6) by using the algebraic product operator defined in (5.9) and the m membership functions and the corresponding fuzzy sets in (6.0):

$$\hat{S} = \bigotimes_{k=1}^{m} \hat{S}_k \Leftrightarrow \hat{\mu}_S\{p|c_1,\cdots,c_m\} = \hat{\mu}_{S_1}\{p|c_1\} \times \cdots \times \hat{\mu}_{S_m}\{p|c_m\} \qquad (6.3)$$

Column IV in Table 1a shows the success rates of the membership function $\hat{\mu}_S\{p|c_1,\cdots,c_m\}$ in (6.3) with respect to the pre-1955 EDF occurrences and Column IV in Table 1b shows the prediction rates with respect to the 1956-1993 EDF occurrences. The two rates are shown as purple lines in Figures 5a and 5b, respectively. Obviously the membership function $\hat{\mu}_S\{p|c_1,\cdots,c_m\}$ in (6.3) completely failed as a prediction model.

The cause of the failure is currently under study, but it appears evident that the multiplication of the membership functions in expression (6.3) requires the assumption that each input layer provides "independent" information. This is obviously not the case in geomorphology.

6.4 *Algebraic sum operator*

We obtain the fifth estimator of the fuzzy set **S** introduced in expression (4.6) by using the algebraic sum operator defined in (5.10) and the m membership functions and the corresponding fuzzy sets in (6.0):

$$\hat{S} = \bigoplus_{k=1}^{m} \hat{S}_k = \left(\bigotimes_{k=1}^{m} \hat{S}_k^{\;c}\right)^c \Leftrightarrow \hat{\mu}_S\{p|c_1,\cdots,c_m\} = \sum_{k=1}^{n}\hat{\mu}_{S_k}\{p|c_k\} - \sum_{k=1}^{n}\sum_{j=k+1}^{n}\hat{\mu}_{S_k}\{p|c_k\}\hat{\mu}_{S_j}\{p|c_j\} + $$

$$+\cdots+(-1)^m \prod_{k=1}^{m}\hat{\mu}_{S_k}\{p|c_k\} = 1 - \prod_{k=1}^{m}\left(1-\hat{\mu}_{S_k}\{p|c_k\}\right) \qquad (6.4)$$

It appears that (6.4) is one of the best fuzzy operators among the all the operators considered in this contribution. The prediction patterns based on the membership function $\hat{\mu}_S\{p|c_1,\cdots,c_m\}$ in (6.4) are shown in Figure 4. The prediction patterns are compared with the pre-1955 EDF landslide occurrences shown in black in Figure 1 to obtain the success rates illustrated in both Table 1a and the corresponding graph in Figure 5a. The prediction patterns shown in Figure 4 are also compared with the 1956-1993 EDF landslide occurrences shown in red in Figure 1 to obtain the prediction rates illustrated in both the Table 1b and in the corresponding graph in Figure 5b.

The pixel values were again divided into 11 classes according to the membership function $\hat{\mu}_S\{p|c_1,\cdots,c_m\}$ in (6.4). The pixels with the highest 5% values among all the pixels in the whole area were classified as the class "0-5%" corresponding to Column V ("algebraic sum") of Row 1 in Table 1a and Column V of Row 1 in Table 1b. They are shown as "purple red" pixels in Figure 4. The pixels with the next highest 5% values were classified as the class "0-10%" corresponding to Column V / Row 2 in Tables 1 and representing "red" pixels in Figure 4. We repeat the classification nine more times and the 11 classes are the ones corresponding to Column V in Tables 1. (fig. 4, see color plate pages, following p. 225)

While Column V / Row 2 in Table 1a shows 0.593, the success rates of the prediction in Figure 4 with respect to the pre-1955 EDF occurrences, Column V in Tables 1b shows 0.410, the prediction rates of Figure 4 with respect to the 1956-1993 EDF landslide occurrences. The 0-10% class (purple red and red) shown in Figure 4 contains 59.3% of the pre-1955 EDF occurrences and 41% of the 1956-1993 EDF occurrences, while this class occupies only 10% of the whole study area. Comparing the previous four values in this row (left hand side), 41% is the best predictor for the EDF landslides to come in the next 38 years (1956-1993).

Comparing the six prediction rates shown in Figure 5b and the corresponding Table 1b, the membership functions from two operators, maximum in (6.2) and this algebraic sum operator (6.4) are two most promising functions as represented by the prediction models for landslide hazard mapping.

6.5 *Gamma operator*

We obtain the sixth estimator of the fuzzy set S introduced in expression (4.6) by using the gamma operator defined in (5.11) and the m membership functions and the corresponding fuzzy sets in (6.0):

$$\hat{S}=f\left(\bigoplus_{k=1}^{m}\hat{S}_k, \bigotimes_{k=1}^{m}\hat{S}_k|\gamma\right) \Leftrightarrow \hat{\mu}_S\{p|c_1,\cdots,c_m\} = \left(\prod_{k=1}^{m}\hat{\mu}_{S_k}\{p|c_k\}\right)^{1-\gamma}\left(1-\prod_{j=1}^{m}\left(1-\hat{\mu}_{S_j}\{p|c_j\}\right)\right)^{\gamma} \quad (6.5)$$

When $\gamma = 0$, (6.5) reduces to (6.3), the algebraic product operator described in Section 6.3. When $\gamma = 1$, (6.5) reduces to (6.4), the algebraic sum operator described in Section 6.4. Hence the gamma operator can be regarded as a generalization of both the algebraic product operator and the algebraic sum operator.

Column VI in Table 1a shows the success rates of the membership function $\hat{\mu}_S\{p|c_1,\cdots,c_m\}$ in (6.5) using $\gamma=0.5$ with respect to the pre-1955 EDF occurrences. Column IV in Table 1b shows the prediction rates with respect to the 1956-1993 EDF occurrences. The two rates are shown as gray lines in Figures 5a and 5b, respectively. As for the algebraic product operator in Section 6.3, the membership function $\hat{\mu}_S\{p|c_1,\cdots,c_m\}$ in (6.5) failed completely as a prediction model. The reason for the failure is identical to the one suggested for the algebraic product operator. (fig. 5, see color plate pages, following p. 225)

6.6 *Landslide counts*

To further interpret both success and prediction rates, we can use the numbers and proportions of individual landslides instead of their area expressed as numbers of pixels corresponding to the different hazard classes. Table 4 shows such rates for the two best predictions, i.e., using the fuzzy minimum and the algebraic sum ($\gamma=1$) operators. There we can see the very high counts of landslides in the higher hazard classes. The values in Table 4 can be compared with the corresponding values in Tables 1a and 1b, Columns II and V.

Table 2a. Aspect map. For each class in each layer, the total number of pixels, the number of pixels affected by pre-1955 EDF landslides within the class, the membership function in (5.5) based on the pre-1955 EDF landslide occurrences, the membership function based on the expert's knowledge and the class description.

Class	Total number of pixels (30x30m)	Number of pixels affected by pre-1955 landslides	Membership function based on pre-1955 landslides	Membership function based on Expert's opinion	Class description
1	56013	1667	0.030	0.050	339°-23°
2	54144	1376	0.025	0.025	24°-68°
3	55366	931	0.017	0.020	69°-113°
4	56663	625	0.011	0.010	114°-158°
5	43095	640	0.015	0.010	159°-203°
6	49638	660	0.013	0.010	204°-248°
7	69136	1382	0.020	0.020	249°-293°
8	67150	2110	0.031	0.025	294°-338°
9	2052	5	0.002	0.000	Flat areas

7 INTRODUCTION OF THE EXPERT'S OPINION

As shown in the column, *"Membership function based on pre-1955 EDF landslides"* of Table 2, the membership functions are based on the past landslide occurrences in (5.5) by computing the frequency distribution of the past (pre-1955) EDF landslide occurrences with respect to each of the supporting "pieces of evidence", i.e. the map layers. The resulting membership functions should be reviewed by experts as we have suggested in (5.6). The modified membership functions by an expert (Luzi, pers. comm.) are shown in the column *"Membership function based on Expert's opinion"* of Table 2. We can repeat all the experiments made in Chapter 6 using the modified membership functions instead of the membership functions based on the past landslides.

Table 2b. Distance from fault map. See also Table 2a caption.

Class	Total number of pixels (30x30m)	Number of pixels affected by the pre-1955 EDF landslides	Membership function based on the pre-1955 EDF landslides	Membership function based on the Expert's opinion	Class description (m)
1	16170	221	0.014	0.05	Fault line
2	4129	70	0.017	0.025	< 30
3	28350	387	0.014	0.012	30 - 90
4	28036	383	0.014	0.005	90 - 150
5	39504	579	0.015	0.005	150 - 300
6	92665	1651	0.018	0.005	300 - 500
7	137167	2612	0.019	0.005	500 - 1000
8	107236	3493	0.033	0.005	> 1000

Table 2c. Distance from drainage map. See also Table 2a caption.

Class	Total number of pixels (30x30m)	Number of pixels affected by the pre-1955 EDF landslides	Membership function based on the pre-1955 EDF landslides	Membership function based on the Expert's opinion	Class description (m)
0	21573	0	0	0	Unmapped area
1	29658	856	0.029	0.050	Drainage line
2	54520	1538	0.028	0.025	< 30
3	48877	1244	0.025	0.012	30 - 60
4	52327	1129	0.022	0.005	60 - 90
5	267875	4629	0.017	0.005	> 90

Table 2d. Slope map. See also Table 2a caption.

Class	Total number of pixels (30x30m)	Number of pixels affected by the pre-1955 EDF landslides	Membership function based on the pre-1955 EDF landslides	Membership function based on the Expert's opinion	Class description
0	21573	0	0	0	Unmapped area
1	90235	464	0.005	0	$0 \leq \theta < 5$
2	43177	1057	0.024	0.01	$5 \leq \theta < 8$
3	44584	2236	0.050	0.02	$8 \leq \theta < 11$
4	42481	2195	0.052	0.03	$11 \leq \theta < 14$
5	37660	1387	0.037	0.04	$14 \leq \theta < 17$
6	38721	899	0.023	0.05	$17 \leq \theta < 20$
7	54973	764	0.014	0.06	$20 \leq \theta < 25$
8	41117	306	0.007	0.07	$25 \leq \theta < 30$
9	32628	82	0.003	0.08	$30 \leq \theta < 35$
10	25059	6	0.000	0.08	$35 \leq \theta < 45$
11	2622	0	0	0.08	$45 \leq \theta \leq 90$

For the Fabriano data, both the membership functions are shown in Table 2. We have applied the algebraic sum operator discussed in Section 6.4 to the modified membership functions and obtained a new fuzzy set and its corresponding membership function:

$$\tilde{\mu}_S\{p|c_1,\cdots,c_m\} = 1 - \prod_{k=1}^{m}\left(1 - \tilde{\mu}_{S_k}\{p|c_k\}\right) \qquad (7.1)$$

where $\tilde{\mu}_{S_k}\{p|c_k\}$ are modified by the expert's opinion (as shown in Tables 2).

Comparing the new results obtained by expression (7.1) with the results obtained by (6.4), and using the original membership functions, the success rates for the pre-1955 EDF landslide data and the prediction rates for the 1956-1993 EDF occurrences from the modified membership functions are shown in Table 3. The table contains the success rates (these are also shown in Column V in Table 1a) and prediction rates (these are also shown in Column V in Table 1b) from Section 6.4, using the original membership functions. The graphs representing the success and prediction rates of Table 3 are shown in Figure 6. There the broken black line is also shown as a red dotted line in Figure 5a, and the continuous thin line in Figure 6 is also shown as a red line in Figure 5b.

Table 2e. Engineering geology map. See also Table 2a caption.

Class	Total number of pixels (30x30m)	Number of pixels affected by the pre-1955 EDF landslides	Membership function based on the pre-1955 EDF landslides	Membership function based on the Expert's opinion	Class description
0	21573	0	0	0	Unmapped area
1	4216	0	0	0	Micritic limestones in thick strata
2	64962	0	0	0	Micritic limestones with chert
3	16365	0	0	0	Marly limestoness
4	70314	0	0	0	Limestones and marly limestones with chert
5	828	27	0.033	0.01	Coarse-grained sandstones in thick strata
6	32759	649	0.020	0.02	Arenaceous-pelitic association
7	832	1	0.001	0.025	Pelitic-arenaceous association
8	1216	12	0.010	0.03	Pelitic unit
9	50698	78	0.002	0.001	Fluvial and fluvial-lacustrine association
10	13325	303	0.023	0.03	Uncemented calcareous debris
11	1141	177	0.155	0.025	Cahotic deposits
12	27884	22	0.001	0.001	Calcareous marls
13	149	0	0	0	Lacustrine deposits and travertines
14	91257	8004	0.088	0.09	Eluvial-colluvial deposits
15	41500	119	0.003	0.001	Cemented calcareous debris
16	35811	0	0	0	Silt and cobbles

Table 2f. Land-use map. See also Table 2a caption.

Class	Total number of pixels (30x30m)	Number of pixels affected by the pre-1955 EDF landslides	Membership function based on the pre-1955 EDF landslides	Membership function based on the Expert's opinion	Class description
0	22596	59	0.003	0	Unmapped area
1	149	0	0	0.01	Absent or scarce vegetation
2	185750	5412	0.029	0.03	Cultivated
3	43150	1071	0.025	0.03	Vineyard
4	66530	1071	0.016	0.02	Grassland or pasture
5	127313	1583	0.012	0	Broadleaf woods
6	27339	195	0.007	0	Conifer woods
7	199	0	0	0	Poplar groves
8	1804	5	0.003	0	Olive groves

As shown in Figure 6 and in the corresponding Table 3, the success rates, illustrated as a broken gray line from the modified membership functions, are a little inferior to the success rates in the broken black line (the red line in Figure 5a) from the original membership functions. The prediction rates shown as a continuous gray line from the modified membership functions are marginally better than the prediction rates shown as a continuous black line (the red line in Figure 5b) from the original membership functions. Hence we can conclude that the expert's knowledge provided only a marginally better prediction model.

The modification of the membership functions by the expert's knowledge is particularly important and necessary when the data base appears to be under-representing the natural setting of the mass movement under consideration.

8 CONCLUDING REMARKS

This contribution has presented prediction models based on the theory of fuzzy sets using five basic fuzzy operators and we have compared their results to the "direct" procedure. The success rates for the training past EDF landslide occurrences (pre-1955 data) and the prediction rates for the future occurrences (1956-1993 data) are shown in Tables 1a and 1b. Figures 5a and 5b illustrate Tables 1a and 1b, respectively, in graphic form. One of the most crucial aspects of a prediction model is a test of the validity of the model. We have achieved this cross validation by time-partitioning of the data base. Another important emphasis that we have made in this contribution is that the prediction model should be for the prediction of future landslide hazard, and not for reproducing the pattern of past EDF landslides. This is the reason why we have introduced the "direct" procedure in Section 4.3., where we have shown that such a direct procedure produced the best possible success rates as shown by Column I of Table 1a and by the corresponding brown line in Figure 5a. The same procedure has badly failed in predicting future 1956-1993 EDF landslides as shown by Column I of Table 1b and by the corresponding brown line in Figure 5b.

Table 3. Success rates for pre-1955 EDF landslide occurrences and prediction rates for 1956-1993 EDF occurrences. Membership functions using pre-1955 data and expert's knowledge and algebraic sum operator described in Section 6.4, Fabriano study area, central Italy. Corresponding graphs are shown in Figure 6.

Area Class (%)	Color	Success rates, fuzzy set, algebraic sum, membership function based on the pre-1955 data	Prediction rates, fuzzy set, algebraic sum, membership function based on the pre-1955 data	Success rates, fuzzy set, algebraic sum, membership function modified by the expert's opinion	Prediction rates, fuzzy set, algebraic sum, membership function modified by the expert's opinion
0.05 0-5	pink	0.362	0.174	0.325	0.253
0.10 0-10	red	0.593	0.410	0.548	0.493
0.15 0-15	orange	0.764	0.632	0.758	0.712
0.20 0-20	yellow	0.880	0.821	0.852	0.828
0.25 0-25	light green	0.919	0.863	0.897	0.863
0.30 0-30	green	0.935	0.897	0.926	0.888
0.35 0-35	blue-green	0.948	0.916	0.944	0.908
0.40 0-40	cyan	0.958	0.936	0.963	0.929
0.45 0-45	light blue	0.966	0.952	0.972	0.947
0.50 0-50	blue	0.975	0.960	0.982	0.951

Even if a specific application area of landslide hazard prediction has been used, general application criteria can be obtained by the comparison. They demonstrate the following points:

1. A spatial data base with all the landslide characteristics including topographic (slope and aspect), geo-technical, geological, and temporal settings needs to be constructed to obtain a prediction.

2. Space-time partitioning of the data base must be done to verify the results of the predictions. The cross-validation test using time-partitioned data should be performed on

Table 4. Success and prediction rates of the classes (5% apart) for two estimation procedures in terms of the corresponding numbers and proportions of individual EDF landslides in the Fabriano study area, central Italy.

Classes	Number of landslides intersected using Fuzzy minimum operator in equation (6.1). (Proportions in brackets) see Fig. 3.		Numbers of landslides intersected using the Fuzzy algebraic sum operator (gamma=1) in equation (6.4) (Proportions in brackets) See Fig. 4	
	"Success rate" Out of 97 landslides which occurred prior to 1955 (compare with Column II of Table 1a)	"Prediction rate" Out of 33 landslides which occurred between 1956 and 1992 (compare with Column II of Table 1b)	"Success rate" Out of 97 landslides which occurred prior to 1955 (compare with Column V of Table 1a)	"Prediction rate" Out of 33 landslides occurred between 1956 and 1992 (compare with Column V of Table 1b)
0-5%	86 (0.887)	25 (0.758)	70 (0.772)	20 (0.606)
0-10%	93 (0.959)	33 (1.0)	92 (0.948)	29 (0.879)
0-15%	93 (0.959)	33 (1.0)	96 (0.990)	31 (0.939)
0-20%	96 (0.990)	33 (1.0)	97 (1.0)	31 (0.939)
0-25%	96 (0.990)	33 (1.0)	97 (1.0)	33 (1.0)
0-30%	96 (0.990)	33 (1.0)	97 (1.0)	33 (1.0)
0-35%	97 (1.0)	33 (1.0)	97 (1.0)	33 (1.0)
0-40%	97 (1.0)	33 (1.0)	97 (1.0)	33 (1.0)
0-45%	97 (1.0)	33 (1.0)	97 (1.0)	33 (1.0)
0-50%	97 (1.0)	33 (1.0)	97 (1.0)	33 (1.0)

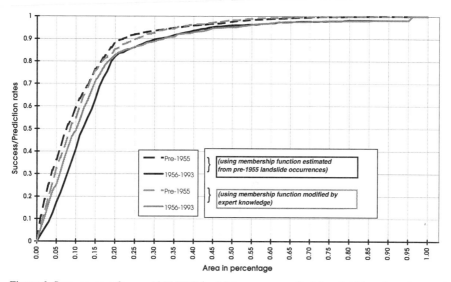

Figure 6. Success rates for pre-1955 EDF landslide occurrences (training data) based on the membership functions using pre-1955 data and the expert's opinion. Prediction rates for 1956-1993 EDF landslide occurrences are based on the membership functions and on the expert's opinion in the Fabriano area, central Italy. The membership functions are integrated by using the fuzzy algebraic sum operator. The corresponding tabular data are shown in Table 3.

the results when future events are predicted using the past events. Such validation test is an essential component of constructing a prediction model, particularly if either the data base does not contain adequate information or where geologic events are poorly understood.

3. As discussed in section 7, the modified membership functions by an expert are shown in the column "*Membership function based on the Expert's opinion*" of Table 2 provided

little additional prediction "power" in this application, as illustrated in Figure 6. Although we have made very little improvement in this study area, the modification is particularly important and necessary when the data base appears to be under-representing the geomorphologic settings of the mass movement. When the data base does not contain "sufficient" information on the past geologic events, the introduction of expert knowledge to construct or modify the distribution of the past occurrences with respect to the input spatial data plays very important role and becomes a necessary tool for fine tuning the prediction.

4. In general, the direct method (Section 4.3) and the algebraic sum operator (Section 6.4) should be applied for the prediction if the data provide reasonable support.

5. Extensive experimentation on the six fuzzy set prediction models should be made in several study areas to strengthen the general situation here identified.

The authors of this contribution have dealt extensively with probabilistic models for landslide hazard zonation (Chung & Fabbri 1998). They are currently studying another quantitative prediction model based on Dempster-Shafer's evidential theory proposed by Chung & Fabbri (1993) as an alternative to the fuzzy set model for geomorphologic hazard zonation studied in this contribution. A model based on that evidential theory is being evaluated in the following terms: (a) the level of knowledge to be introduced in the data base; (b) the evaluation of the belief in the evidence; and (c) the semantics of representation of natural and man-induced processes.

ACKNOWLEDGMENTS

We wish to thank David. F. Garson who has helped us to prepare all figures and tables in this paper, and Andy N. Rencz (Geological Survey of Canada), H. Kojima (Science University of Tokyo) and R. S. Divi (CNS Geomatics Inc.) who have read an earlier draft of the manuscript and have provided many useful comments. We gratefully acknowledge the cooperation of Lucia Luzi, of the Italian National Research Council in Milano who has provided us not only with the Fabriano data but also with the expert's opinion in Table 2. The study was partly funded by a Japan Science Technology Fund granted to the Geological Survey of Canada through the Canadian Department of Foreign Affairs. We also wish to acknowledge a NATO Collaborative Research Grant Awarded to the authors in November 1997.

REFERENCES

Agterberg, F.P., Bonham-Carter, G.F. & Wright, D.F. 1990. Statistical pattern integration for mineral exploration. In, G. Gaal & D.F. Merriam (eds), *Computer Applications in Resource Estimation, Prediction and Assessment of Metals and Petroleum*: 1-21. New York: Pergamon Press.

Aronoff, S. 1989. *Geographic Information Systems: A Management Perspective*. Ottawa: WDL Publications.

Bickel, P.J. & Doksum, K.A. 1977. *Mathematical Statistics: Basic ideas and selected topics*. San Francisco: Holden-Day Inc.

Chung, C.F. & Fabbri, A.G. 1993. The representation of geoscience information for data integration. *Nonrenewable Resources*, 2(2): 122-139.

Chung, C.F. & Fabbri, A.G. 1998. Probabilistic prediction models for landslide hazard mapping. Photogrametric Engineering and Remote Sensing, in press.

Chung, C.F., Fabbri, A.G. & van Westen C.J. 1995. Multivariate regression analysis for landslide hazard zonation. In A. Carrara & F. Guzzetti (eds), *Geographical Information Systems in Assessing Natural Hazards*. Dordrecht: Kluwer Academic Publ.: 107-133.

Chung, C.F. & Leclerc, Y. 1994. A quantitative technique for zoning landslide hazard. *Papers and Extended Abstracts for Technical Programs of IAMG'94, 1994 International Association for Mathematical Geology Annual Conference, October 3-5, 1994*: 87-93. Mont Tremblant, Quebec.

Carrara, A. 1983. Multivariate models for landslide hazard evaluation. *Mathematical Geology* 15(3): 403-427.

Carrara, A., Cardinali, M. & Guzzetti, F. 1992. Uncertainty in assessing landslide hazard and risk. *ITC Journal* 1992(2): 172-183.

Fabbri, A.G. & Chung, C.F. 1996. Predictive spatial data analysis in the geosciences. In M. Fisher, H.J. Scholten, & D. Unwin (eds), *Spatial Analytical Perspectives on GIS in the Environmental and Socio-Economic Sciences*: 147-159. London: Taylor & Francis.

Leclerc, Y. 1994. The Design of FM: Data Integrating Package for Zoning Natural Hazards in the Developing Countries. Unpublished M.E. Des. Thesis, Environmental Science, Faculty of Environmental Design, University of Calgary, Canada.

Leclerc, Y. & Chung, C.F. 1993. FavMod for integration of spatial geoscience information. Calgary: Geological Survey of Canada, Open File 2577 (Software).

Luzi L., 1995, GIS for Slope Stability Zonation in the Fabriano Area, Central Italy. Unpublished M.Sc. Thesis, ITC, Enschede, The Netherlands.

Luzi, L. & Fabbri, A.G. 1995. Application of favourability modeling to zoning of landslide hazard in the Fabriano area, central Italy. Proc. *Proceedings of Joint European Conference and Exhibition on Geographic Information, JEC-GI'95: "From Research Application Through Cooperation", Den Haag, The Netherlands, March 26-31*, 1: 398-403.

Spiegelhalter, D.J. 1986. A statistical view of uncertainty in expert systems. In W.A. Gale (ed), *Artificial Intelligence and Statistics*: 17-55. Reading, Massachussetts: Addison-Wesley Publ. Co.

Zadeh, L.A. 1965. Fuzzy sets. *IEEE Informatics and Control*, 8: 338-353.

Zadeh, L.A. 1968. Probability measures of fuzzy events. *Jour. of Math. Analysis and Application*, 10: 421-427.

Zadeh, L.A. 1978. Fuzzy sets as a basis of a theory of possibility. *Fuzzy Sets and Systems*, 1: 3-28.

Impacts of land-recycling projects; Evaluation and visualisation of risk assessments

K. Heinrich
Swiss Federal Institute of Technology, Ecotechnics and Sanitary Engineering, Lausanne, Switzerland

ABSTRACT: This paper, based on land recycling approaches in Germany and the Netherlands, discusses the differences between EIA's for brownfield rehabilitation and conventional EIA approaches. Because of the complex relationship between land-recycling and the natural and socio-economic environment, EIA for site remediation includes informational gaps and uncertainties. The handling of uncertainties by means of a Fuzzy Expert System is illustrated for the assessment of the environmental status quo of a site. Examples from mining industry and wastewater purification illustrate specific components of EIA in land-recycling. Two cases studies to illustrate the application of the concepts discussed are presented; one deals with the design of a land recycling plan for a polluted site and another one with the actual rehabilitation of a mine spoil heap.

1 INTRODUCTION

In urban agglomerations with high population density, virgin land is rare - particularly in cases where industrial and commercial acquisitions require vast development areas. Derelict industrial sites are often contaminated and due to ecological, socio-economic and political reasons such "brownfield" areas have to be remediated. Land-recycling or brownfield rehabilitation remediates soil and groundwater pollution, while simultaneously reducing the development pressure on virgin land (greenfields). The greenfields would be taken out of a more or less functioning environmental context elsewhere, to be consumed as building lots. At polluted sites the re-establishment of the soil functions is of utter importance, since soil forms a significant border system of litho-, hydro-, atmo- and biosphere, that pervade and influence one another. Hence, EIA for risk assessment and remediation planning should focus on the process of removal, remediation or decontamination of soil and groundwater to the levels of local background values for prevailing chemical species. EIA have to determine the risks involved for people (death, injury, disease), goods (property damage, economic loss) and the environment (loss of flora and fauna, loss of amenity) in all stages of the land-recycling process: before re-mediation (present situation, i.e. status quo), during remediation (implementation phase) and after implementation (monitoring of measures). In the ideal case, the site is prepared for any further use (multi-functionality) after remediation. From this point onwards any envisaged project has to pass through a conventional EIA procedure .

Due to the heterogeneous and hazardous nature of brownfields, it is essential to adjust and integrate conventional EIA approaches, geotechnical routines and the good practices of site investigation to the specific and exceptional boundary conditions in land-recycling. Consequently risk assessment and remediation planning differs from EIA for a new structure on greenfields. In a greenfield case of a planned chemical plant, for instance, the future potential (negative) impacts have to be assessed and minimised. In a rehabilitation project, the present situation must be changed and it is expected to have a strong future impact on the environment, namely a positive one. Land-recycling is a curing process in which the present contamination situation is the determining factor rather than potential negative environmental effects in the future and their prevention or reduction. This explains why the site's environmental "status quo" has such a dominant role in brownfield EIA. The case of a new chemical plant planned in the neighbourhood of a groundwater protection zone may serve as an example: If an EIA would trigger rigid governmental restrictions and expensive preventive measures, the plant may not be cost effective and an alternative location may become more opportune. A soil or groundwater contamination in the vicinity of a protected aquifer provokes an immediate reaction with a high remediation priority. Instead of imposing restrictions, EIA and remediation planning are ruled by the contamination situation from the very beginning of a land-recycling project. Hence, EIA is rather applied to assess the environmental impact of unavoidable remediation measures during implementation, operation and monitoring phase, without having a zero-option. The EIA has to balance and minimise expected environmental impacts of the implementation of remediation measures to reach remediation success. However, this EIA process is governed by the remediation requirements, which are determined by intensity and type of contamination as overruling site factor. Site factors are a combination of both geo-factors and the dominant anthropogenic impacts, caused by the industrial use of the terrain. The ground at industrial sites typically can be classified as anthropogenic soil or "urbic anthrosol" (FAO-UNESCO 1988). Physical properties of urbic anthrosols are usually varying in plan and depth over small distances. This characteristic heterogeneity and the resulting lack of data/information provoke uncertainties about the stratigraphy as well as quantities, distribution, mobility and possible emissions of contamination. This causes frequent and unexpected changes in the layout of remediation planning, implementation and the following treatment of contaminated material. Good practice in Germany and the Netherlands is a parallel and iterative planning strategy, a round table with all parties involved. Heinrich (1996) and ASTM (D 5730, 1998) state that iterative remediation planning strategies enable a constant discussion and a flexible adaptation to unexpected new findings during risk assessment and remediation process. This flexible approach may lead to an adaptation of the future land-use to remediation requirements. The iterative EIA/remediation planning can lead to considerable financial and practical profit, because it covers design, remediation efficiency and economy, physical impact estimations and the assessment of indirect economic and social effects of the project and each adjustment of it. What has been said above about parallel and iterative planning strategies has also been mentioned by Durucan (1995) in the context of EIA.

Due to the unpredictable conditions at brownfields, it becomes obvious that the present situation is not a precise reference point in time but rather a highly dynamic time period that may persist until the end of the remediation implementation. This dynamic component is not new to the conventional EIA concept, starting from the present situation, it assesses the future development of a site without the implementation of a certain action (zero-option). However, EIA deals with gradual mid and long term influences, instead of sudden day-to-day changes typical for site remediation. Thus, the 'dynamic' status quo is much more important and complex in brownfield rehabilitation than in greenfield projects. The present state of a site is a function of all past events at the site and the surrounding environment and is therefore very important in land-recycling.

Point of departure in risk assessment is a desk study. It includes a profound analysis of the site's contamination history. This paper introduces and discusses the Multi-Temporal Desk Study (MTDS). The MTDS as the most economical part of a risk assessment becomes increasingly important, since the growing number of site remediation cases demand more accurate prescience and more cost-effective approaches for the assessment of contamination distribution in soil and groundwater. Even though the base for decision is widened constantly by incoming data from the progressing assessment and remediation process, obviously a major deficiency in land-recycling remains, that is imprecision, uncertainty, vagueness and heuristic approaches. Among other factors the heterogeneous character of the urbic anthrosols and the small-scale variations of pollution is responsible for this. In the sequel we will cast a light upon a method for handling these inherent difficulties in modelling site factors during risk assessment by means of a Fuzzy Expert System called SAFES. Uncertainty prevails throughout the whole EIA/risk assessment/remediation process. To exemplify we discuss the handling of fuzziness for the assessment of the present situation (contamination) of a wastewater purification plant. In order to familiarise the reader with some problems of geomorphology and land-recycling specific aspects will be reviewed in an example of a mine tailing heap. Both sections are based upon the following section as a theoretical discussion about the steps of land-recycling analysis.

2 STEPS IN LAND-RECYCLING ANALYSES

Due to complex interlocking of geo-factors and anthropogenic aspects, planning of site investigation and remediation measures on derelict and contaminated sites is a demanding task for a multi-disciplinary team. Therefore, the entire planning process must be adopted and structured with special regard on its intrinsic complexity. Genske & Thein (1994) recommend a three-step-approach for cost effective remediation of derelict sites. Figure 1 shows a modified four-step-version of this model.

2.1 *Risk assessment (present situation)*

Risk assessment can be regarded as a tool for project management, in the sense that it allows decision-makers to approve, redesign or reject a remediation measure. Replacing the word "remediation measure" with the word "project" in the last sentence we find back a definition of EIA (Sadler 1988). If a severe contamination is encountered it is not the question if the rehabilitation is realised, in fact the question is only "how" and "which" measures must be taken. Therefore, risk assessments investigate site factors, that is, geo-factors and anthropogenic impacts that caused a certain present state of con- tamination. As an inventory, it is an indispensable pre-requisite for a successful site re mediation. Similar to conventional EIA, risk assessments should attempt to foresee and provide against hazards and difficulties that may arise during realisation of remediation measures and the subsequent construction of infrastructure and buildings. The assessed site factors constitute basic design assumptions for remediation and spatial and constructional planning on the site. In cases such as open pit mining, quarries, tailing heaps and subsidence areas, geomorphologic properties belong to the most prominent site factors that affect the local or regional geo-ecosphere.

2.1.1 *Phases of risk assessment*

The risk assessment and EIA comprises two phases: an indirect approach, called desk study and a direct exploration, called field reconnaissance. In the sequel, both phases of the risk assessment are discussed in more detail.
Risk assessments basically must reveal the following properties/ hazards of a terrain:
1. Geomorphologic properties and processes: Degradation and erosion (emissions by water transport and deflation), accumulation (immission by sedimentation), weathering (e.g., pyrite weathering and sulphate release), pH, EH, solubility, architectural landscape modelling in the course of rehabilitation, etc.
2. Geotechnical properties: Quality of foundation soil, buried foundations and obstructions (e.g., sewers, gas pipes, tanks), slope stability, grain size distribution (e.g., clay content, related to mobile behaviour of contaminants).
3. Geological properties: Lithology and structure, e.g. stratigraphy, discontinuities, fracturing (preferential flow) .

4. Hydro-geological and hydrological properties (transport and fate of contaminants): Hydro-geological units, phreatic aquifers, hydraulic connections, groundwater flow, permeability (e.g. aquitards, aquicludes), location of existing wells (sampling groundwater up- and downgradient), precipitation, infiltration capacity (e.g., surface seal), seepage, surface runoff.
5. Physical/chemical properties of pollutants: Toxicity, mobilisation potential, volatilisation - hazardous soil gas emission (explosion and intoxication hazard), spatial distribution, etc.
6. Biological properties: Macro and micro flora and fauna (diversity, protected species, potential for phyto-remediation), microbiology (potential for natural attenuation, bio-remediation, metabolism), bio-turbation of soil (preferential flow).
7. Hazard-pathway-target relations: Exposition and resulting health or environmental risk (field crew, site users, neighbours, biosphere).
8. Technical and use history: Analyses of facilities, production engineering, functional and constructional changes for deduction of contamination potentials.
9. Socio-economic aspects: Regional development potential, spatial planning, city planning, working places, image improvement, investment incentives.

2.1.2 Desk study/MTDS

Desk studies help to define the scope of EIA/risk assessment by means of object recognition, site boundaries, land ownership, current and past land use, natural site characteristics, eco-functional units and any other relevant information for remediation planning. Desk studies provide a preliminary conceptual site model that serves as representation of the site's environmental "status quo". With the progressing assessment this model becomes successively more substantial.

The desk study is a non-destructive "sampling" method without any environmental impact on the site. Most of the time it relies on soft or indirect data/and information, such as aerial photos, maps or site plans. As a first judgement desk studies consider the development of a potentially contaminated site by assessing changes of anthropogenic and environmental state over time to obtain indicators of potential soil contamination. This section is an introduction to the method of historical analyses and the Multi-Temporal Desk Study – MTDS in particular. MTDS uses maps and airphotos for interpretation and subsequent evaluation. The information gained by this method is related to other sources of information, for instance the analyses of the site's history in literature, reports, etc., and walk-over surveys. Especially regarding chemical hazards, the results of the desk study provide essential information for the protection of the field crew and future occupants of the site. A desk study must consist of:
1. Reconstruction of all former manufacturing plants and any other features (e.g. geomorphology: tips, excavations, cut and fills, subsidence areas, filled bomb craters, landslides, piling pattern/techniques of tips, landfills, etc.).
2. Reconstruction of land-use zones (railway lines, roads, dumping sites, etc.).
3. Synthesis and interpretation of the information, including an impact assessment.
4. Zoning of potential soil contamination.
5. Modelling, visual presentation and production of a report.

The MTDS is a non-destructive and indirect method, where no drillings are conducted. Only surface features are interpreted in order to obtain a model of the site's contamination history. Normally a MTDS is meant to be the preparatory stage of a proper field campaign. The stock of information includes both, the different editions of topographical and thematic maps and aerial photos of different years of survey.

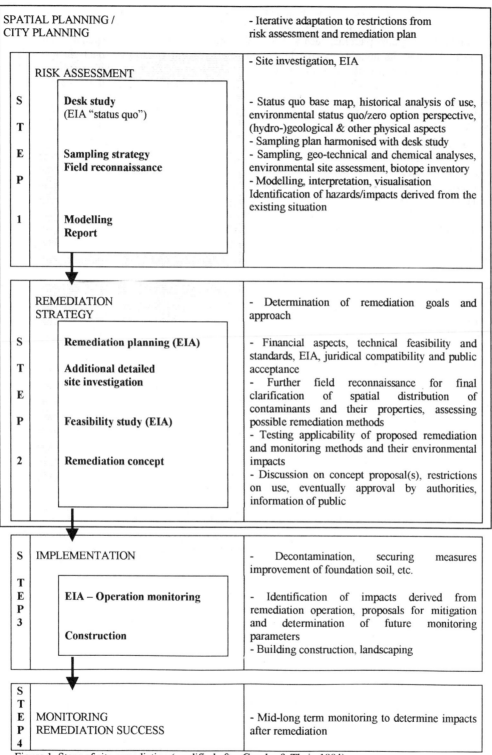

Figure 1. Steps of site remediation (modified after Genske & Thein 1994)

MTDS' cannot directly detect concentrations of pollutants in the underground. The course of site pollution and the contamination potential of the locations can only be deduced by an interpretation of spatial patterns, such as use of the site and its installations, spatial layout and shape of features, its mutual spatial relation and the temporal spatio-functional changes. To assign contamination potentials to features (e.g., installation, specific areas) the interpreter must combine and interpret the visual information. Thus, it is important to consider the characteristics of the information sources. For example, geological, geomorphologic, hydro-geological and pedological maps represent a generalised picture of the (sub)-surface properties only, because strata and hydrological boundaries, positions of structural features and soil units are often interpolated. Bearing in mind the relative small size of many derelict and polluted sites (often only a few hectares), the information drawn out of the maps basically reveals overviews and indicates tendencies. Under ideal conditions only, surface reflection, texture, drainage pattern or vegetation cover observed on aerial photos can be matched with geological and pedological units in maps. However, the undifferentiated area map sign for urbic anthrosols (usually hatching in black and white) often blurs the map information.

Nevertheless, maps are useful sources of information, especially with regard to the development of bigger manmade geomorphologic features, such as mine tailings, subsidence lakes and quarries and open pit mines. Furthermore, they provide greater spatial overviews, which are relevant for potential mobile behaviour of contaminants (e.g. groundwater up and down gradient). Shifted pollutants may eventually be encountered and even accumulated in the vicinity of the original hot spot, or in the deeper underground beneath an anthropogenic layer. Especially the topographic maps facilitate a time- sequential approach and the documentation of spatio-functional changes of industrial sites, because of their frequent updates in different editions that may even reach back until the beginning of industrialisation. They also show the development of manmade geomorphologic features, such as the progressing exploitation of a quarry or the growth of a tip and their consequences for hydrologic patterns. An essential component is the documentation of site's technical history and its subsequent merging with the information of the other parts of the MTDS. Information on this can be retrieved from site plans, libraries, (company) archives, municipalities and old newspapers and former staff. This step reveals insight in industrial production practices and technological changes over the years, the use, type and quantities of materials and resources, the output of by-products and waste, waste treatment practices, accidents, damages during war, etc.

Aerial photo interpretation is very suitable for the EIA/risk assessment in connection with the development of geomorphologic features. MTDS is especially appropriate for large scale items such as erosion gullies and related accumulation patterns (transport and fate), excavations, bomb craters, fillings of smaller hollow forms (dumping sites) or solid forms (tips, dumping site), leachate contaminated springs and small drainage paths such as ditches. The later may indicate potential surface water sampling locations. The deduction of material quality (reflection characteristics) and the quantities (volume calculations) is important for the assessment of the remediation effort to be made; i.e. the type and dangerousness of the material expected to be handled and treated. The type and state of vegetation cover gives clues on erosion susceptibility and soil contamination for example (vegetative stress, necrotic or eutrophic peculiarities). Furthermore, slope stability can be estimated and sinkholes detected. Sinkholes may emerge due to underground voids because of old infrastructure, such as cellars, etc. This matters for working safety reasons and foundation soil consideration. Colliery mine tailings often show a special phenomenon: spontaneous ignition due to the presence of coal and methane gases within the heap. These hot spots can be detected by infrared airphotos.

In case of subsidence, aerial photos can reveal the change in relief directly by measurement or indirectly by means of ecological changes, for instance the transition towards a wetland. In both cases airphotos show potential and restrictions for future use.

The reflective characteristics of an object are influenced by the type/material and by its physical state. Differences in contrast or similarities can generally be used for object classification, such as material distinction. Material distinction is performed by differences or similarities of the object's tone, colour and texture; i.e., its visual impression of coarseness or smoothness. The texture is produced by characteristics, which are too small to be detected individually (e.g., grains, tree leaves). Similar materials show similar textures on the same photo. Fundamentals of environmental remote sensing are extensively discussed in Vincent (1997).

Regular coverage with airborne photos started in the 1920s and 30's; since the late 1930's, stereographic images are available in the Netherlands and Germany. Unlike maps, photos represent a most complex and undifferentiated snapshot of reality at a certain point in time and the visible features are un-selected, un-coded and not generalised. A set of photos gives only a sequential view on the site with considerable gaps in-between (usually 1-5 years). Furthermore, distortions can be significant and have to be equalised by geo-referencing. This can easily be done in the raster domain of GIS. The interpretation of aerial photos results in a set of maps of the respective survey years, that later is combined with the remaining MTDS information to complete the picture.

During MTDS the connection of identified objects with (technical) background knowledge is essential in order to assess the contamination potential of a certain location or a specific installation. This is a process of interpretation that always has a subjective component. The differentiation and identification of certain objects can only be done after an analysis and comparison of distinctive criteria, a logical association of indicators and a reasoning process reflecting human heuristics as strategy for problem solving or rule of thumb approach. A skilful interpretation can result in additional information even about invisible characteristics such as groundwater (Nefedov & Popova 1978) or soil pollution (Borries 1992). In this case, visible image elements serve as a registration base for invisible features, which are related causally and spatio-functionally with the invisible items. Section 3 gives full particulars on the handling of imprecision and subjectivity in MTDS and EIA.

2.1.3 Field Reconnaissance

Field reconnaissance for environmental site characterisation represents a direct approach to risk assessment. The field campaign should delineate essential prevailing features qualitatively and quantitatively, including interactions and synergism. This goal does not differ from conventional EIA. Remediation planning should identify and locate, both horizontally and vertically, significant soil and rock masses and groundwater conditions present within a given area, and establish in conjunction with the desk study the site factors (ASTM 1998). Soil, soil gas and groundwater samples are collected and analysed, the foundation soil is investigated, trial pits and trenches are excavated, etc. If no rigid sampling grid must be applied because of legislation, the sampling plan should be harmonised with desk study results in order to reduce costs for sampling and analyses.

For the evaluation of soil and groundwater samples and the subsequent manner of spatial modelling, standards differ from country to country. As an example of evaluation system for the physico-chemical properties of contaminated sites, this paper refers to the Dutch system of target/intervention values (WBB 1995).

The target values define the standard for 'clean' soil and groundwater. The intervention values define the concentrations for chemical species and compounds in soil and groundwater, which are considered to impair the functional properties of the soil for humans, flora and fauna. Exceeding intervention values indicate severe soil contamination. *In-between target values and intervention values lies an interval of intermediate values. If a concentration exceeds 1/2 × (intervention value + target value) a detailed site investigation has to be performed. A detailed list of parameters and values is provided in Dutch Soil Protection Act (WBB 1995). The* reference base for defining these values are the natural background values in the Netherlands. Among other aspects, the proper environment and the exposure of pollutants are also taken into account. Thus, the standards are flexible and can be adapted to specific situations. To speak about severe cases of contamination at least 25 m^3 of soil or 100 m^3 of water must exceed the concentration above the intervention values. Variables such as clay and organic contents can be introduced by means of a Soil Type Correction Formula. Equation (1) is an example of the formula for inorganic compounds:

$$I_b = I_{st} \left(\frac{A + B \times \% \text{ clay} + C \times \% \text{ organic matter}}{A + B \times 25 + C \times 10} \right) \qquad (1)$$

where
Ib = intervention values, valid for soil to be evaluated in [mg/kg]
Ist = intervention values, valid for standard soil/sediment (10% organic matter and 25% clay) in [mg/kg]
% clay = mean % of clay in soil to be evaluated (% of weight ignition loss in relation to total dry weight of soil)
% organic matter = mean % of organic matter in soil to be evaluated (% of weight of minerals < 2μm in relation to total dry weight of soil)
A,B,C = constants related to the compounds As, Ba, Cd, Cr, Co, Cu, Hg, Mo, Ni, Zn (see WBB 1995).

The Dutch strategy of risk assessment distinguishes three consecutive modules of site investigation, based upon target and intervention values and the degree of homogeneity or heterogeneity of the contamination distribution (Hardeveld et al. 1995).

2.1.4 Environmental impacts of field reconnaissance

Sampling for risk assessments and remediation planing can be divided in destructive a non-destructive methods. Some geophysical methods such as ground radar, EM and geomagnetic surveys belong to the non-destructive and indirect sampling methods. Consequently, their environmental impact during sampling is negligible. Seismic methods has successfully been applied for detection of landfill extensions (Vogelsang 1993). However, they should be applied with care on derelict industrial sites, since the vibrations may trigger the collapse of underground voids such as cellars, old mining shafts, etc. Thus, it is a potentially destructive method that may cause an immediate impact on the surface and is a potential risk to the field crew.

Direct sampling is an essential part of site investigation for environmental purposes. Drilling is the most frequent method to collect soil and rock samples for "in-situ" characterisation of physical and geo-hydrological properties, chemical and biological characteristics, subsurface lithology, stratigraphy and structure, as well as hydro-geological units. An alternative to drillings are trial pits. Direct sampling can cause two significant environmental impacts, which have to be assessed before and during the sampling campaign. One problem is related to the handling of the (contaminated) excavated material. Especially trial pits produce considerable quantities of material. The excavation material has to be dumped and perhaps treated on-site or off-site. The sampling crew may need to wear protection cloth, perhaps transport has to be organised, possibly regarding certain safety regulations, such as sally ports with cleaning facilities and coverage of the transported material or special transport vehicles. During on-site storage of excavated material, a cover from rain and wind should prevent discharge. Groundwater extraction for sampling, pumping and tracer tests also produces wastewater that has to be handled adequately. Soil gas extraction provides with samples of volatile fractions.

A different problem may be a legal one: by excavation, the material may change its legal status and become hazardous waste instead of remaining contaminated soil. Hence, different regulations may apply.

A further possible severe impact on the environment can result from the perforation of the different geological layers by drillings. This creates highly permeable vertical connections between potentially polluted and non-polluted layers or aquifers. In addition, cross contamination may lead to erroneous analytical results. Drilling in industrial sites has to be conducted with care, because service lines such as electricity and gas (explosion) may cause accidents when damaged.

2.2. Remediation planning

For brownfield rehabilitation the step of remediation planning has to assess the different remediation approaches and their impact on the environment. Often one has to deal with a combination of methods to reach a certain remediation goal.

The examples in Figure 2 illustrate such a combination of excavation and thermal treatment. The first step of the process is an excavation of the contaminated material. Afterwards thermal treatment is applied. Incineration transfers pollutants from one medium (soil/soil water) towards the exhaust gas, which has to undergo an after treatment. The process and the environmental impacts are depicted in Figure 2 by a sketch. The next assessment aspect within the chart applies to the type of contaminants that can be successfully remediated by this method. Under the last point the efficiency and the environmental impacts are specified. This chart allows for an easy comparison of the different remediation methods and also gives insight into a possible combination of methods as the evaluation of the thermal treatment demonstrates. It is not in the scope of this paper to address all types of remediation techniques, their possible combination and their impact upon the environment. LaGrega et al. (1994), and EPA (1996) give comprehensive overview of remediation techniques and strategies.

3 UNCERTAINTY AND FUZZINESS LAND-RECYCLING

As mentioned above, uncertainty and imprecision are typical characteristics of environmental geo-science. Often enough information exists but in a form that can only be accessed heuristically by means of interpretation and experience of the expert rather than in a computational adequate manner. However, lack of information, ignorance, gradual transitions and even contradictory information can be handled in a mathematical way, based upon the fuzzy set theory (FST) and fuzzy logic. Zadeh (1965) has introduced the notion of fuzziness, as a means for representing and manipulating non-stochastic, i.e. non-random uncertainty. Additionally to the hard analytical data of the 1-or-0 type, FST enables the introduction of soft information into a computer model. It can be seen as a further replenishment of aspects in the complex context of environmental relations. This paper focuses on aspects of fuzziness in land-recycling and discusses a geo-scientific application in form of a Soil Assessment Fuzzy Expert System (SAFES), that has been tested at a contaminated purification plant in the Netherlands.

Rommelfanger (1988) distinguishes three types of fuzziness: intrinsic, informational and relational fuzziness. The discussion here refers to applications to EIA within the framework of risk assessments and to the "status quo" assessment in particular.

The intrinsic fuzziness can be understood as heuristic fuzziness. Heuristic fuzziness stands for evaluation and solution strategies which are based upon intuitions, experience, educated guesses and rule of thumb approaches. Usually it is reflected and handled by the use of linguistic descriptors. To this category belong linguistic descriptions such as 'high degree of pollution', 'strong exposure', 'high risk of (contaminated) dust deflation', 'very low intake rate', 'reasonable costs', etc. Even if a crisp (or precise) critical value exists, such as the Dutch intervention values, the objectivity suggested is actually non-existent. Although critical values for soil contamination are based on scientific research, the determination of the final levels of pollutants is basically a somewhat arbitrary political decision about what is considered to be contaminated and hazardous.The strained relation between legal directives, geo-scientific necessities and cost effectiveness determines the frame of decision making in land-recycling.

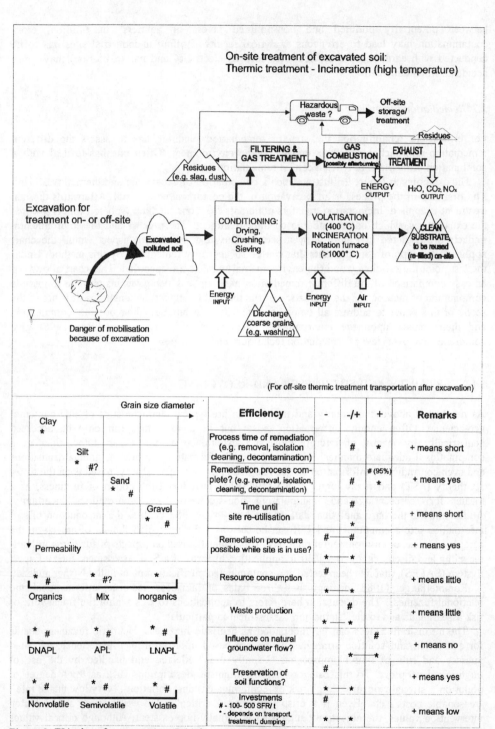

Figure 2. EIA-chart for excavation & high temperature thermic treatment (lecture notes Swiss Federal Institute of Technology 1999); Symbols: * = excavation, # = thermic treatment. Abbreviations: DNAPL – Dense Non-Aqueous Phase Liquid, APL - Aqueous Phase Liquid, LNAPL – Light Non-Aqueous Phase Liquid, SFR = Swiss Francs.

The next paragraph shows some examples of intrinsic fuzziness, being part of the decisions in land-recycling:
- Gradual transition from background values to increased pollutant concentrations.
- Imprecise relation between concentration in the soil, exposure, intake rate and actual resorption in the human body.
- The interpretative character of the evaluation (e.g., use of experience, educated guesses, rule of thumb, inter- and extrapolations).
- Uncertainty about alternatives and preferences (e.g., different scenarios).
- Uncertainty about consequences (e.g., metabolisms during bio-degradation).

Informational fuzziness is well known in environmental science. In EIA/risk assessments, cost effectiveness requires usually a compromise between the absolute necessary and the desirable investigation effort. That usually causes a lack of information/data concerning soil samples and chemical analyses. This often may be considered as being insufficient from the scientific perspective, whereas the legal aspects are satisfied. The borehole information has to be inter- and extrapolated to fill the informational gaps between the drilling sites. Semantically this is often expressed by the notion 'potential', for instance, 'high contamination potential'. Some typical examples are:
- Indistinct and ambiguous features (e.g., multi-temporal aerial photo interpretation)
- Incomplete information (e.g., historical analysis, one-dimensional borehole data, toxicological knowledge)
- Ignorance and overseeing of features, mistakes in site investigation approach (e.g., sampling strategy and method)
- Mistakes during sampling (e.g., improper closing of sampling containers [volatilisation], wrong storage)
- Limited precision of chemical analyses, (e.g., mixed samples, improper analyses).

Relational fuzziness is a combination of either intrinsic and/or informational fuzziness. For example: 1. 'Is the pH about 4 or lower and the clay content is not very high, then Cd can easily be mobilised'. 2. 'Is the degree of contamination high, then the image of a site becomes worse and the incentives for investments are very low'. The relational fuzziness is typical for the decision-making process and complexity in site remediation. Some examples for typical problems are mentioned below:
- Increasing complexity (e.g., increasing toxicological knowledge, environmental consciousness, multi-disciplinary planning approaches)
- Varying boundary conditions at different sites (e.g., varying exposures towards features to be protected, future use)
- Overlapping and synergetic effects (e.g., relation between solubility of some heavy metals and pH, and subsequent mobilisation).

These examples of fuzziness in rehabilitation projects lead to the conclusion that all three types of fuzziness are inherent parts of land-recycling, especially in decision making. Partial ignorance, errors, uncertainties, vagueness, imprecision, ambiguity and semantic heuristic descriptors are inherent to evaluation in risk assessment.

For desk studies, Genske et al. (1993) mention that the application of high precision CAD or GIS becomes relative, when considering the inaccuracy of the basic evaluation material, because of its age and/or the former less advanced production techniques. Distortions of historical maps and aerial photos must be equalised and geo-referenced. Siehl (1993) states for 3-D GIS that automation and the subsequent need for mathematical precise or crisp determination of a problem and its solution may end in impoverishment of the information content and in premature restrictions in the possible solution space. In consequence, the final results may be unrealistic.

It s not in the scope of this chapter to explain the mathematical background of fuzzy set theory and fuzzy logic. A good introduction is presented by Zimmermann (1991); the application of FST for MTDS is explained in detail in Heinrich (1998).

4 THE USE OF FUZZY EXPERT SYSTEMS

Nowadays, processing and interpretation of data in environmental science are often done using expert systems (ES). Expert system developing tools offer the opportunity to create special purpose systems adapted to specific objectives and syntheses of knowledge from various disciplines. Bichteler (1986) gives the following definition: 'An expert system is an intelligent computer program which uses knowledge and inference procedures to model human expert reasoning in a narrow and specialised field'. Hall & Kandel (1992) define a fuzzy expert system (FES) as a system 'which incorporates (next to the properties mentioned above) fuzzy sets and/or fuzzy logic into the reasoning process and knowledge representation scheme'.

Besides the urgent necessity to find better and economic ways of assessing site factors, such as geomorphologic processes or transport and fate of pollutants, the idea for implementing fuzzy logic into modelling of site factors was triggered by the inherent fuzziness in risk assessments and EIA. Crisp ES' are already valuable tools in geo-science and land-recycling, because of their support to decision making. However, FES' can help to bridge the typical gaps caused by the different inherent types of fuzziness. Generally, FES' help to get better results from the evaluation processes. Especially the natural language interface, the knowledge management potential, the ability of weighing uncertainties and incomplete data and the possibility to incorporate human heuristics make FES' useful tools for decision-making in land-recycling (Heinrich 1999).

4.1 *Knowledge-Based Geographic Information Systems*

The combination of ES with GIS offers the opportunity to integrate spatial information and heuristics. This solves the crucial deficiency of GIS' of barely reflecting heuristic logic in its mathematical modelling and numerical data analysis techniques. The binary mathematical "modus operandi" in a GIS, which allows only rigid Boolean functions, causes difficulties in handling imprecision. This dilemma can be solved by means of combining (F)ES and GIS (Özmutlu 1995). This merging of systems is subsumed under the abbreviation KBGIS (Knowledge-Based Geographic Information System).

KBGIS' are made up of a GIS with its spatial data base management and the knowledge base organising the expert knowledge. The GIS is responsible for data input, query, storage, retrieval, spatial analysis and data output. The (F)ES provides the control function of the entire system by managing the knowledge base and the inference operations. In addition, the ES regulates the data flow between (F)ES and GIS and vice versa, and supplies rule-based inference. Subsystems can be integrated in case algorithms and/or other spatial operations are required. Running a KBGIS will reduce the data management time and increase research and application efficiency (Özmutlu 1995).

5. CASE STUDY: FUZZY ENVIRONMENTAL ASSESSMENT

A SAFES-prototype (Soil Assessment Fuzzy Expert System) was tested for the environmental assessment of the "status quo" at a purification plant for industrial wastewater. This example shows all the typical fuzzy elements of any EIA or remediation planing. As stated earlier these comprise the intrinsic, informational and relational fuzziness, which are an inherent part of any decision base in environmental assessment. In case of MTDS, decisions about the location of future sampling points have to be made, whereas during remediation planing the decision for or against certain techniques is the goal of decision making. The preliminary risk assessment and the MTDS for the purification plant serve also as an example of the coupling between SAFES, a GIS for spatial modelling and a CAD for visual refinement of the model.

The evaluation of the spatial distribution and the degree of contamination potentials was based upon a multi-temporal map and aerial photo interpretation, which included 11 aerial photos over a time span of 29 years (1966-1995) as well as different editions of topographical, geomorphologic, geological, hydro-geological and pedological maps. The fuzzy contamination potentials and the overall real world peak-soil contamination were compared by means of GIS models. A distinction between different contamination parameters was not made, since the fuzzy model of contamination potentials depicts the chance to encounter soil/groundwater pollution only. However, the whole approach suffers from some deficiencies: because of external limitations (lack of time) and in contradiction to the normal site investigation approach, this desk study was not carried out before the proper field campaign. Therefore, not all 52 drilling sites were optimally placed. In consequence, the comparability between real world model and fuzzy model is sub-optimal. The SAFES model represented in Figure 3 is derived from the main features of the site, that is tanks and pipes shown on the aerial photo of 1995. Potential contamination, that may have been caused in the past and that possibly may still persist in the ground is disregarded in this comparison. In addition, geo-hydrological information from maps was not yet integrated into the fuzzy model. Possible directional mobile behaviour can be interpreted indirectly only from a conventional groundwater model and the resulting gradients.

The comparison of real world and fuzzy model has been performed in a relative manner. The overall real world peak-soil contamination and the fuzzy potentials have been normalised to the closed interval [0,1]. The purpose of the comparison was to find out parallels between both models regarding the trends in spatial distribution and the degree of soil contamination.

Figure 3a depicts the fuzzy contamination potentials determined by means of SAFES for 1995. Two mayor peaks of contamination potentials (no.1 and 2) in the north and in the western central part can be seen. Furthermore, two smaller peaks (no. 3 and 4) are visible in the central and southern part of the terrain. It is obvious that the fuzzy model underestimates the actually observed and modelled soil contamination.

The comparison with the overall peak contamination in the real world model of Figure 3b reveals some striking parallels. The four mentioned peaks of the fuzzy model are also found in the real world model. West of peak 1´, which has about the same magnitude as peak 1, is a second peak (1^a´). Furthermore, the degree of soil contamination encountered at peak 2´ is considerably lower than predicted in the fuzzy model and instead a second peak (2^a´) in the S shows almost the same magnitude. A similar phenomenon occurs at peaks 4´ and 4a´, where peak 4´ is a bit lower than the predicted fuzzy contamination potential of peak 4. In summary, all the four potentially contaminated locations in the fuzzy prediction were found to be contaminated in the real world model, some of them, however, with a different degree of contamination. Moreover, a characteristic doubling of the peaks has been observed in 3 cases. The model of the groundwater surface in Figure 3c reveals one possible explanation for these double peaks. This groundwater model is based upon 48 measures. However, especially in the NE but also in the SW, no readings are available. Yet, it can be assumed that this hydraulic head continues for some distance to the E. The tag 1' points out the spatial coincidence with the lower peak of soil contamination indicated in Figure 3a, as well as the labels 2' and 4' do for the corresponding peaks in the real world model. Cross checks with the types of contaminants observed in the samples belonging to these very peaks, show the presence of potentially mobile pollutants. Since the observed double pinnacles of the real world model are found in the downstream direction of the hydraulic heads, the conclusion can be drawn that the double heads of pollutants concentration in the soil are due to a shift by groundwater. This interpretation is supported by the fact that the very small (fuzzy) peak no. 3 indicated in Figure 3b, is the only one that has no characteristic double peak. Even though this peak is located at the eastern slope

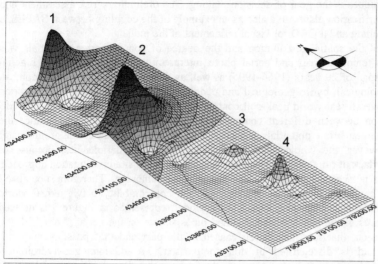

Fig.: 3a

Interpolated model of fuzzy contamination potentials (note slightly changing scales of the models)

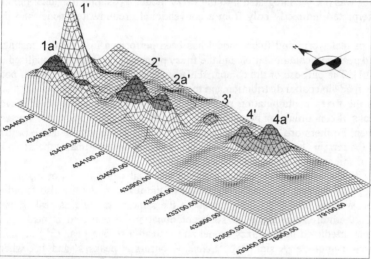

Fig.: 3b

Interpolated real world model of soil contamination based upon 52 drillings (note slightly changing scales of the models)

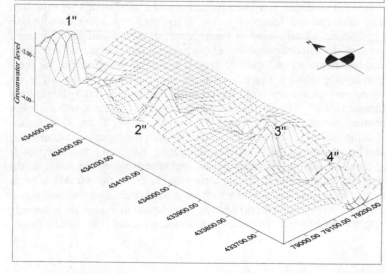

Fig.: 3c

Model of groundwater surface; NE and SW of the site no data available (note slightly changing scales of the models)

of a hydraulic head, this pinnacle is special; it is lying in a drip pan of a huge tank, which has a horizontal impermeable membrane to avoid the spread of contaminants into deeper layers after possible (and observed) spillage. Hence, this potential contamination is isolated from the site's geo-hydrological system and immobile. This comparison shows that the results of a fuzzy MTDS modelling with SAFES produced promising results. Even though the example only reflects one aspect of EIA and risk assessment, there is a big potential of a successful transfer of this fuzzy approach to other parts of the site remediation planning and other environmental problems. However, SAFES still has to be improved. In this context, especially the dominant object orientation must be transferred into a two dimensional and later three-dimensional approach. That is, geological and hydro-geological aspects have to be introduced to refine SAFES. Moreover, fuzzy interpolation techniques should be introduced in the ES. In a later stage, a combination with crisp techniques (stochastic) would be desirable.

6. CASE STUDY: LAND-RECYCLING OF A MINE TAILING HEAP

This second case study further illustrates the concepts presented above, applied to geo-morphological rehabilitation.

In the sections above, some of the most important reasons for land-recycling have been mentioned. A further significant reason for land-recycling is the image improvement of affected regions. This means for instance, the creation of pleasant natural surroundings, a good cultural infrastructure or - in general terms - a high quality of life. Hence, next to the positive environmental impacts a 'cleaned-up' and positive image is also attained by brownfield rehabilitation. A good visible and prominent example in the context of geomorphology is the land-recycling of mine tailing heaps for recreational purposes and/ or as natural areas. The general rehabilitation goals for mine tailing heaps are an environmentally friendly re-mediations followed by re- naturalisation and conversion towards a "*Landschaftsbauwerk*" (landscape structure), i.e., the integration of a huge anthropogenic geomorphologic body into the landscape. These landscape integrated tailing heaps (Fig. 4) are characterised by smooth silhouettes, echelon slopes and benches, sweeping layouts, flattening slope edges and smooth domes to create a more natural appearance. Thus, for the envisaged recreational use and the ecological conservation it is necessary to take advantage of the heap's attributes. The *Halde Preussen* heap, formed over 30 years ago serves as an example of conversion of a heap into a "*Landschaftsbauwerk*".

The tip is located SE of the city of Lünen/ Germany; it has a volume of about 850,000m^3 and is situated on the site of a former colliery (hard coal) and coking plant. Beside dead rock, a mixture of coal/sewage sludge and flotation residues has been dumped. Subordinated components are waste, rubble, wood, rubber and other mining wastes.

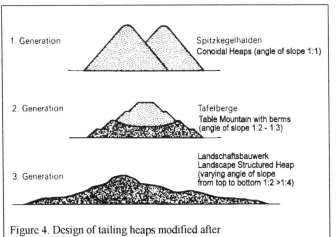

Figure 4. Design of tailing heaps modified after Bergbauhandbuch (1994) and Olschowy (1993)

6.1 Environmental impacts of tailing heaps

The mining process can be divided into four distinct phases: Exploration development and construction, production and decommissioning and post production. (Sinding 1998). Land-recycling is a process of the post production phase. Usually the previous phases have either been performed without any EMS (Environmental Management System) or only with systems that do not match current standards. In the course of the formerly haphazard growth of tailing heaps in mining regions, far-reaching consequences occurred for the environment. Besides the obvious changes of relief, tailing heaps influence soil composition, water balance, micro and meso-climate as well as the biosphere. This affects also their surroundings. EIA for the present state of heaps recognised negative impact of heaps with special regard to soil balance, water balance but also vegetation, landscape and the whole image of the neighbouring municipalities.

For instance, the soil balance is influenced because of the coverage of vast areas with dead rock that destroys large surfaces with pedologically intact soils. In consequence, important soil functions are impaired. Due to changes in chemistry and soil water contents, the soil balance and edaphon may be disturbed. Before the growth of vegetation, the heap was subject to erosion and deflation, affecting the tip's vicinity. The water balance is influenced, because of deterioration of hydrological self-regulation. Surface water system stability is affected by the loss of retention capacities. The self-cleaning capacity is reduced and effects on evapotranspiration and recharge of groundwater also occur. As soon as surface waters are destroyed, the climate regulation function is affected, too. Chemical changes are likely, because of chloride and sulphate inputs. There is a negative impact on land uses such as forestry, agriculture or water resource management, aquatic biotopes and recreational areas (D'Alleux 1991).

Tailing heaps are extreme habitats for vegetation, especially with respect to physical and chemical site factors and to the dynamics of weathering and subsequent changes in soil parameters. Important and special properties of tailing heaps in the Ruhr District are:
- Strong relief variation, high differences in elevation, steep slopes (1:1.3 at old tips)
- Low content of fine grained material before weathering
- Lack of nutrients
- Low field capacity (soil moisture retained by capillarity)
- Strong acidification because of pyrite weathering
- High vulnerability to erosion (disturbance in soil development)
- Xero-thermic micro climate (high surface temperatures plus lack of water)
- Strong winds causing deflation of fine grains/detritus/humus and soil desiccation.

Because of special site factors, in particular hydro-geological conditions *Halde Preussen* evolved to an unusual ecological island for this region. Surface waters in adjacent ditches are strongly saliniferous due to weathering leachates. The trench's watercolour is brownish, because of high ferro-hydroxide content and white manganese hydroxide precipitation. At this place a biotope with a halophile plant association (*Bolboschoenetum maritima*), which is considered to be rare in this region, developed.

6.2 Heap facies

The heap facies can be defined as a combination of ecological, geological, geomorphologic, technical and economical processes and characteristics. The heap's facies controls the anthropogenically-induced geological processes on and in the heap. For instance, the intensity of ferrous sulphide concentration (FeS, FeS_2) is a function of the lithofacies of the operated seams (e.g. pyrite content, grain size), the dead rock portion, the way of excavation and intensity of pyrite weathering (Fig. 5). The weathering process is basically divided in 3 steps (Schöpel &

Thein 1991). In the first phase easily soluble minerals, mainly NaCl, and water-soluble sulphates are mobilised. In that short first period the leachate springs are highly mineralised. The second phase is characterised by $NaHCO_3$ (natron hydrogen carbonate) and Na_2SO_4 (natron sulphate) leachate caused by sulphide oxidation. In the next stage exhaustion of the acid-buffering system and a subsequent pH reduction to levels lower than 4, can cause mobilisation that enriches effluents with heavy metals (Mn, Fe, Zn, Cd, Cu, Pb, Cr and partly As).

Besides leachates other aspects had to be considered during the rehabilitation process. The following characteristics of the heap facies had to be taken into consideration and were iteratively integrated in the EIA for the remediation planning:
- Spontaneous ignition, due to high coal and methane content and unsystematic heap layout and management in the past.
- Methane and smoke emission, dust and noise emissions.
- Soil, ground and surface water contamination.
- Weathering (physical, chemical) causing leachate of acids, sulphates, sulphides, aluminium and heavy metals polluting surface groundwater, furthermore erosion and deflation.
- Special pedologic aspects (acidification, mineralisation).
- Foundation soil problems regarding future structures (restaurant, look-out tower).
- Slope stability problems, ((re-)design of berms, dead rock disposal techniques, angle of slopes, soil-protective vegetation).
- Change of hydrological patterns.
- Specific heap vegetation with species and associations under protection.
- Influence on micro- and meso-climate, hydrology, etc.

The heap facies and rehabilitation are also related to spatial planning, socio-economic and legal boundary conditions, such as:
- Landscape architectural harmonisation with surroundings in general and neighbouring main recreational area in particular.
- Legal problems, financing, management, acquisition.

Under this caption an EIA was performed as a part of desk study, risk assessment, remediation assessment and feasibility study. Table 1 below lists the main EIA items.

Table1. Main items of the EIA at "Halde Preussen"

EAI components	Investigated fields	
Risk assessment, site remediation planing:	Pollution: Hydro-geology:	• Soil and groundwater contamination • Transport and fate of pollutants • Soil gas and methane emissions
	Foundation soil: Geomorphology: Ignition situation:	• Slope stability • Weathering, erosion, deflation, accumulation • Localisation of hotspots
Ecological assessment	Biotope-types Biotope network Balancing	• Plant associations • Establishment of links • Buffer zone
Spatial planning assessment	Recreation Landscaping Infrastructure	• Public parkland • "Landschaftsbauwerk", landmark • Parking, access roads, recreation facility
Socio-economic assessment	Image improvement Working places	• Pleasant natural surroundings, cultural infrastructure, better quality of life

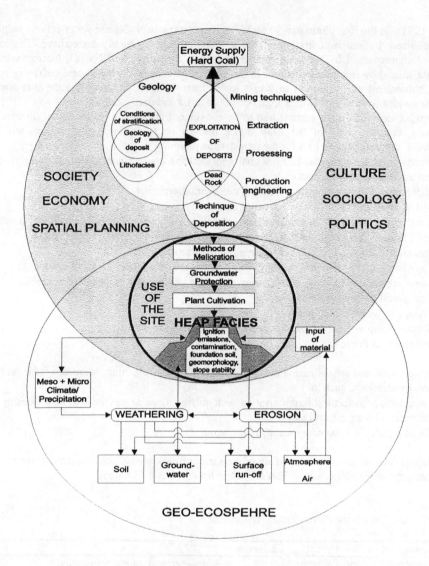

Figure 5. Control factors of heap facies and environmental impact of the heap on the surroundings (modified after Kerth & Wiggering 1991).

6.3 *Goals of heap rehabilitation*

During the EIA/risk assessment inquiry of the "status quo" it became obvious that the heap formed a unique ecological entity that evolved as it lay 'derelict'. The EIA pointed out that this resulted from particular physical site factors created through pedologic, climatic, hydrological and hydro-geological conditions brought about by the mining activities and subsequent dereliction. As the most appropriate goal the EIA formulated the heap's integration into the landscape as buffer zone and substitute for neighbouring areas under mining exploitation or endangered by inadequate use. From the point of landscape architecture, an aesthetic harmonic blending of the tailing heap with its surrounding landscape was considered to be desirable. The ecological aim was the conservation of vegetal succession areas within reservation zones, which meant the preservation of much of the tailing heap's morphology and the partly unique vegetation. Furthermore, the re-naturalised heap including the foreland, would establish a new link with the regional biotope network. In accordance to the EIA, the rehabilitation and recreation approach had to be chosen. However, this goal turned out to be contradictory, since it meant the heap's preservation in terms of its present ecological state on the one hand, and radical changes such as levelling and establishing new vegetation for public recreational parkland on the other hand. A different goal was to establish and control/remediate the present and potential damage the heap caused or could cause through any toxic emissions, such as contaminant leachate, especially as the area would become a public recreational facility. The public access demanded improvements on foundation soil and slope stability. The rehabilitation had the following goals:

- Environmental goals: ecological buffer area, ground-water protection, soil decontamination, etc., and additionally conservation of valuable biotopes, integration within the surrounding landscape and biotope network
- Safety for visitors, complete reduction of emissions, slope stability, control of ignitions, image improvement of the region.

6.4 Remediation Implementation

In the case of *Halde Preussen*, time pressure for implementation of the projects was high. Therefore, the sub-optimal removal of material was applied for coal and sewage sludge volumes that were contaminated with organic substances and partly with heavy metals. Their treatment was performed off-site.

The groundwater regime was only partly known and the suspicion of pollutant migration towards a bathing lake remained. The lake was planned to be part of the recreational area. As the bathing lake is a most sensitive site for protection, the EIA recommended maintaining groundwater monitoring for another two years. Although surface water should neither be used as drinking nor as industrial water, chemical and exposition analyses showed there was no direct risk for humans beings nor the environment. Today the lake is fully in use as a public bathing lake.

Three of the discovered ignition hotspots had to be treated either by excavation, sealing, and/or compacting of the surface layers. Again, these measures had to be given higher priority than the conservation of valuable biotopes. This was especially true for the locality of the warmest hotspot ($\pm 300°$ C) in the heap. In this zone, the ignition source was sealed with inert material to cut off the O_2 supply from the surface. Another combustion zone was excavated in order to extinguish the low temperature combustion and re-fill the material. To reduce the risk of new spontaneous ignition the material was compressed and basal and lateral contacts to other material were isolated. Because the material was prone to strong C_6H_6 emissions in the past, a porous and coarse grained capping layer was provided to allow gas discharge. Monitoring of the development of potential future heap ignitions will be maintained. Slope stability also had to be improved. This was simply achieved by flattening the slope gradients from 1:1.5 to 1:2.

Planning reality appears to belie part of the formulated EIA goals, namely to assign highest priority to ecology. In the course of the remediation works valuable biotopes, originally meant to be preserved were at least partly destroyed. For compensation, plantations were established in the heap foreland and at other localities in the Ruhr District. However, one essential ecological goal could not be fulfilled: the conservation of the succession areas had to be sacrificed for the sake of soil and groundwater remediation and fire extinction. The more general goal of developing a balance area for ecological valuable sites under pressure remained untouched.

7 CONCLUSIONS

Throughout the paper it has been shown that EIA is an integral and indispensable part of any land-recycling project. The term EIA is usually not directly mentioned in brownfield rehabilitation, since the terminology has been coined in the framework of site remediation. Nevertheless, both risk assessment and remediation planning comprise strong EIA elements. Since site remediation is a curing (re-)action, one difference with conventional EIA is the predominance of the contamination situation and the "dynamic status quo" due to typical uncertainties provoking sudden unexpected findings and irremediable compulsions. Rehabilitation projects in which geomorphological features are essential are usually found in the context of mineral exploitation, mining and quarrying. Especially the case of the tailing heap shows that desirable EIA goals sometimes are sacrificed because of remediation demands. This shows the crucial presence of uncertainty, vagueness and ignorance within brownfield rehabilitation. A better integration of prevailing fuzzy facts may help to avoid wrong decisions and improve the general perception of a contaminated site. A different aspect became apparent in both case studies: sub-optimal practical solutions, either in planing of the field campaign or remediation measures are inherent to site investigation and land recycling.

REFERENCES

ASTM, *D 5730-96*, 1998. American society for testing materials, ASTM standards related to phase II environmental site assessment process: 179. West Conshohocken, PA. USA.

Bichteler, J. 1986. Expert systems for geoscience information processing. Proc. *3rd International Conference on Geoscience Information*. Vol. 1: 179-191. Glenside. Australia. Aust. Miner. Found.

Borries, H.-W. 1992. *Altlastenerfassung und Erstbewertung durch multitemporale Karten- und Luftbildauswertung*. Würzburg. Germany. Vogel Buchverlag.

D'Aalleux, J. 1991. Nutzungskonflikte bei Aufschüttungen von Bergehalden.In Wiggering, H., Kerth, K. (eds).*Bergehalden des Steinkohlenbergbaus*: 59-64. Wiesbaden. Germany. Viehweg Verlag..

Das Bergbauhandbuch. 1994. Wirtschaftsvereinigung Berbau e.V., Bonn (eds). Essen.Germany. VGE Essen

Durcan, S. 1995. Environmental simulation and impact assessment system for the mining industry. Oral presentation, CANMET-EC-USBM workshop 1995. Denver. Colorado. USA.

EPA 1996. EPA environmental engineering source book.. Russel Boulding, J. (ed.). 2 vol. Chelsea. Michigan. USA. Ann Arbour Press..

FAO-Unesco 1988. *Soil map of the world*. Rome. Italy. FAO.

Genske, D., Kappernagel, Th., & Thein, J. 1993. Visualisierung umweltgeologischer Probleme am Beispiel der Reaktivierung von Industriebrachen. In Proc. *145. Hauptversammlung der deutschen Geologischen Gesellschaft*. Krefeld. Germany. Not published.

Genske, D., & Thein, J. 1994. Recycling Derelict Land. In Proc. *1. International Congress on Environmental Geotechniques*, 10.-15.July 1994: 33-43. Richmond. Canada. BiTech Publishers.

Hall, L. O. & Kandel, A. 1992: The evolution from Expert Systems to Fuzzy Expert Systems. In Kandel, A. (ed.). *Fuzzy Expert Systems*: 4-19. London. UK. CRC Press.

Hardeveld, van, W., Moet, D., Smits, A. Vroomen, L.H.M., Oudheusden, van, K., & Timmermanns, F. 1995. Soil sustainability: Dutch policy. In W.L. van den Brink, R.Bosman & F.Arendt (eds). *Contaminated Soil '95:Proceedings of the fith international FZK/TNO conference on contaminated Soil, Maastricht*: 1605-1619. Dortrecht.The Netherlands. Kluwer Academic Publishers.

Heinrich, K. 1996. Brachflächenrecycling - eine ingenieurgeologische Herausforderung. *Proceedings of the "10. Nationale Tagung Ingenieurgeologie".*, *Freiberg*: 196. Essen. Germany. VGE Essen.

Heinrich, K. 1998. Fuzzy modelling and visualisation for risk assessments in land recycling projects. Proc. of the Fifth International Symposium on Environmental Issues and Waste Management in Energy and Mineral Production - SWEMP'98. Ankara.Turkey: 123-133. Rotterdam. The Netherlands. Balkema.

Heinrich, K. 1999. Fuzzy Assessment of Contamination Potentials. PhD-thesis. Delft. The Netherlands. in print.

Kerth, M., & Wiggering, H. 1991. Steinkohlebergehalden als anthropogen-geologische Körper. Wiggering, H., Kerth, K. (Eds). *Bergehalden des Steinkohlenbergbaus:* 47-58. Wiesbaden. Germany. Viehweg Verlag.

LaGrega, M.D., Buckingham, & P., Evans, J.C. 1994. Hazardous waste management. New York. McGraw-Hill.

Nefedov, K., & Popova, T. 1978. *Deciphering of Groundwater from Aerial Photographs.* New Dehli. Amerring.

Olschowy, G 1993: *Bergbau und Landschaft:* 215. Hamburg/Berlin.Germany. Paul Parey.

Özmutlu, S. 1995. *Expert System Approach for Preparation of Special Purpose Engineering Geological Maps.* M.Sc.-thesis. International Institute for Aerospace Survey and Earth Science. Delft. Netherlands. ITC.

Rommelfanger, H. 1988. *Entscheiden bei Unschärfe, Fuzzy Decision Support-Systeme.* Berlin Heidelberg. Germany. Springer Verlag.

Sadler, B.1988. The evaluation of assessment: Post-EIS research and process development. In: Wathern, P. (ed.) *Environmental Impact Assessment. theory and Practice.* London, UK. Routledge.

Schöpel, M., & Thein, J. 1991: Stoffaustrag aus Bergehalden. Bergehalden des Steinkohlenbergbaus. Wiggering, H., Kerth, K.(Eds). *Bergehalden des Steinkohlenbergbaus:* 115-128. Wiesbaden. Germany. Viehweg Verlag.

Siehl, A. 1993. Interaktive geometrische Modellierung geologischer Flächen und Körper. In *Die Geowissenschaften-Forschung und Praxis.* 11: 342-345. Berlin. Germany. Ernst&Sohn.

Sinding, K. 1998. Environmental Impact Assessment and Management in the Mining Industry. Proceedings of the Fifth International Symposium on Environmental Issues and Waste Management In Energy And Mineral Production - SWEMP'98, Ankara/Turkey 1998: 81-86. Rotterdam. The Netherlands. Balkema.

Vincent, K. 1997. Fundamentals of geological and environmental remote sensing. NJ. USA Prentice-Hall.

Vogelsang, D. 1993. Geophysik an Altlasten: 100ff. Berlin. Germany. Springer Verlag.

WBB- Leidraad bodembescherming.1995: Den Haag. The Netherlands. Sdu Uitgeverij.

Zadeh, L. 1965. Fuzzy sets. In *Information and Control.* 8: 383-353

Zimmermann, H. J. 1991. *Fuzzy set theory and its applications.* Dordrecht. Netherlands. Kluwer Academic Pub.

Geomorphology and Environmental Impact Assessment: A case study in Moena (Dolomites - Italy)

M. Marchetti & M. Panizza
Dipartimento di Scienze della Terra, Università degli Studi di Modena, Italy

ABSTRACT: Moena is a small town at the intersection of three valleys (Fiemme, Fassa and San Pellegrino) in the Italian Dolomites (North-Eastern Alps).
   Due to the large number of holiday makers using the town's infrastructures both in summer and winter, the local authorities have acknowledged the need for the construction of a ring-road round Moena for the last twenty years.
   This paper analyses the relationships between landforms, in particular geomorphological assets, and the project of the road, and between processes and project, in order to assess impacts and risks and suggest the mitigation measures that should reduce them.
   Impacts are assessed by means of a geomorphological asset map deduced from the general geomorphological map and combined with considerations on the fragility and value of the geomorphological asset.
   Risks are assessed by means of a geomorphological hazard map deduced from the general geomorphological map and combined with considerations on the vulnerability of the project and other human properties.
   The research points out that impacts on the landforms are quite low whereas some risks are possible in the slope system crossed by the road. In fact, on the left slope of the Fassa Valley south of Moena some small slope processes are possible but their dimensions and amount of movement can be mitigated by diminishing the vulnerability of the road.

1 INTRODUCTION

This work is part of a more general project funded by the European Union regarding Geomorphology and Environmental Impact Assessment (EIA). In particular, in the course of a project entitled: "Geomorphology and Environmental Impact Assessment: a network of researchers in the European Community" funded in the framework of the European Programme "Human Capital and Mobility (HC&M)", the studies aim to evaluate the suitability of a project for assessing the quality of the Environment and to develop a procedure for geomorphological impact assessment of projects on the environment (Marchetti et al. 1995, Panizza et al. 1996a).
   The relations between the geomorphological components (landforms and processes) and the project are considered. The project under examination concerns the construction of a new road around the village of Moena in the Italian Dolomites, North-East of Trento town (Fig. 1). Due to the large number of tourists making use of the resort's infrastructures both in summer and winter, the local administration has for many years recognised the need to construct a new ring-road round Moena.
   Environmental studies in Alpine areas are considered more problematic than in other areas. This is due both to the extreme sensitivity of the Alpine territory and the high value that this environment has for Man, owing to its scenic beauty. In fact, the continuous increase of tourist activities has on the one hand made people aware of the high scientific value of the Environment and its fragility and on the other hand made them aware of the increase of its economic value and the cost for the conservation of the environmental elements.

2 GEOGRAPHICAL SETTING

The study area is the Moena valley, located at the intersection of three valleys: Val di Fiemme, Val di Fassa and Valle San Pellegrino; the latter two valleys end into Moena, where the first begins (Fig. 1).

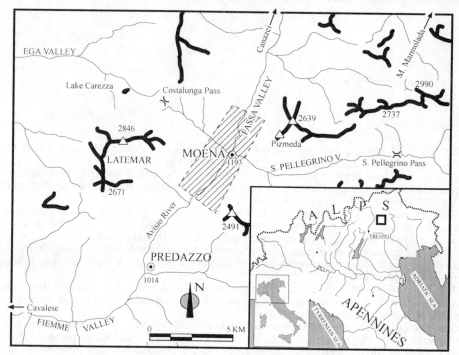

Figure 1. Geographical setting of the study area.

Moena (2600 inhabitants) is set in the Trentino Dolomites, 1184 m above sea level North-East of Trento and is topographically represented in the 1:25,000 scale Italian Military Geographic Institute (Istituto Geografico Militare Italiano, I.G.M.I.) map 11 III SE "MOENA". The highest peaks are below the perennial snow limit; they are: Cima Viezzena (2490 m) SE of the main inhabited centre, Dosso Mezzogiorno (2301 m) South-East of the same village and Mount. Pizmeda (2200 m) to the North-East. The Rio San Pellegrino, running through the valley of the same name, and the Rio Costalunga, on the right-hand side of the main valley, flow into the River Avisio near the village of Moena (Fig. 2). All these rivers are characterised by a permanent flow rate regime which is related both to snowmelt water and precipitations.

In the study area also an artificial reservoir (Lake Soraga) is found North of Moena, which was obtained by damming a narrow gorge eroded through time by the Avisio River and which is now used for power production.

Val di Fassa stretches from South to North between Moena and Canazei; the latter is the most important tourist centre of the area, owing to both the high standard of its tourist resorts and facilities and its proximity to one of the most famous peaks of the Alps: Mount Marmolada (3342 m).

From Moena, Val di Fiemme slopes down to the South towards Predazzo and Cavalese which are two other large and important inhabited centres of the Dolomites.

The climate of the area is characterised by cold, dry winters and cool, rainy summers (Dbw from Köppen classification, although noticeably conditioned by the altitude). The valley floor is affected by snow fall from November to March, whereas at higher altitudes the snow period lasts until May.

Summer is, on the contrary, characterised by short and intense rainfall, occurring practically every day. The mean annual temperature is 5.1°C and the mean annual precipitation is 1107.2 mm.

3 GEOLOGICAL SETTING

The geological formations outcropping in the area are sedimentary and volcanic in origin

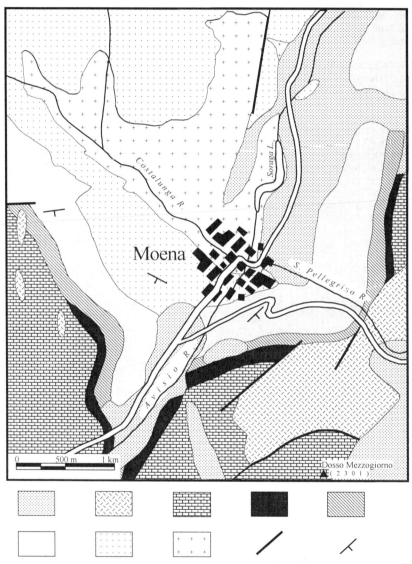

Figure 2. Geological map of the area around Moena. Legend: 1) Quaternary deposits; 2) Volcanic formations (Ladinian and Carnian); 3) "Latemar Limestone" and "Marmolada Limestone" (Ladinian); 4) "Livinallongo Formation" (Lower Ladinian); 5) "Richtofen Conglomerate" and "Dark Morbiac Limestone" (Anisian), "Moena Formation" and "Contrin Formation" (Upper Anisian); 6) "Werfen Formation" (Schytian); 7) "Bellerophon Formation" (Upper Permian) and "Val Garden Sandstone" (Permian); 8) Ignimbrites of the Atesino volcanic complex (Lower Permian); 9) Fault; 10) Bedding attitude.

and belong to the Permian-Triassic sequence of the western Dolomites (Bosellini et al. 1977, 1996, Bosellini, 1996) (Fig. 2).

The oldest outcropping rocks are ignimbrites belonging to the upper unit of the "Complesso Vulcanico Atesino" (Lower Permian). They largely outcrop North of Moena, along the course of the Rio Costalunga and are followed in stratigraphic sequence by the "Arenarie di Val Gardena" (Middle Permian) which crop out in a portion of territory parallel to the Rio Costalunga and North-West of Moena. This formation is made up of coarse reddish sandstones and micaceous siltites, especially in correspondence with the boundary with the overlying "Formazione a Bellerophon". The Bellerophon Formation (Upper Permian), made up of hollow dolomitic rocks alternated with bituminous limestones, is characterised by outcrops limited to the area West of Moena. Since this formation is morphologically similar to the underlying "Arenarie di Val Gardena", in Figure 2 it is associated with the latter. The "Formazione di Werfen" (Schytian) is made up of alternances of grey marly limestones, micaceous siltites, red and yellow calcarenites with oolithic levels and Gasteropoda. It crops out to the South-West and South-East of Moena, along stretches of the Avisio and San Pellegrino Rivers. The "Conglomerato di Richtofen", the "Calcari scuri di Morbiac" (Anisian) and the "Formazione di Moena" and "Formazione di Contrin" (Upper Anisian) outcrop in stratigraphical sequence with limited extensions which are observable on the northern face of Dosso Mezzogiorno. The "Formazione di Livinallongo" (Lower Ladinian) is made up of nodular micritic limestones and turbitic-carbonatic sequences. It outcrops with varying thickness on the North-West face of Dosso Mezzogiorno and on the slopes South-West of Moena. The "Calcari del Latemar" (stratified) and "Calcari della Marmolada" (massive) (Ladinian) close up the sedimentary sequence of the Moena valley. They crop out on the highest slopes South and South-West of Moena. The sedimentary sequence is cut through or is in contact with the igneous rocks of the "Complesso Magmatico Predazzo-Monzoni" and the Ladinian-Carnian submarine lava flows.

Two North-East - South-West trending displacement systems, particularly noticeable on the South-West slope of Mount Pizmeda and the northern one of Dosso Mezzogiorno, affect the whole area. North-West of the Soraga Lake a North - South trending displacement brings the Permian formations of "Arenarie di Val Gardena" into contact with the porphyries.

Most of these displacements are buried under thick detrital covers.

4 GEOMORPHOLOGICAL SETTING

The area investigated is characterised by vast detrital covers of different nature and age. In fact, the geological bedrock outcrops almost exclusively in correspondence with fluvial erosion forms or man excavations (Fig. 3).

The Holocene fluvial and lacustrine deposits are located in the lower portion of the valley floor and are spread over very large extensions. The valley-bottom deposits in the area South of Moena seem to be derived from the damming of the River Avisio valley by large alluvial fans, South of the present study area (Carton & Panizza 1983). A well exposed alluvial fan is found North-North-West of Moena; it has increased its volume owing to the action of the Rio Costalunga and is partially overlapping a vast debris fan placed more to the South. The alluvial fan is made up of rounded boulders with clasts and scanty silty-clayey matrix. The thickness of the deposits is around some tens of metres.

Palustrine deposits are presently found North-West of Moena, on the right-hand slope of the River Avisio valley. The marshy areas occupy the lowest zones of glacial deposits, where counterslopes and ridges make the water runoff difficult. Moreover, the consequent deposition of silts and clays has increased the imperviousness of the terrain.

The glacial deposits bear witness to several glacial expansions, among which the one found East-North-East of the Soraga Lake, at 1380 m above sea level, is the highest in altitude and is considered as the product of the Last Glacial Maximum (Carton & Panizza 1983). Other deposits, found at lower altitudes, are the evidence of lateral and sometimes frontal accumulation due to stadial episodes (morainic arc reaching the Soraga Lake, at 1220 m). In all these cases, the higher hydraulic conductivity of the glacial deposits with respect to the surrounding rock types causes, especially North of Moena, on the right slope of the valley, the formation of stagnant waters.

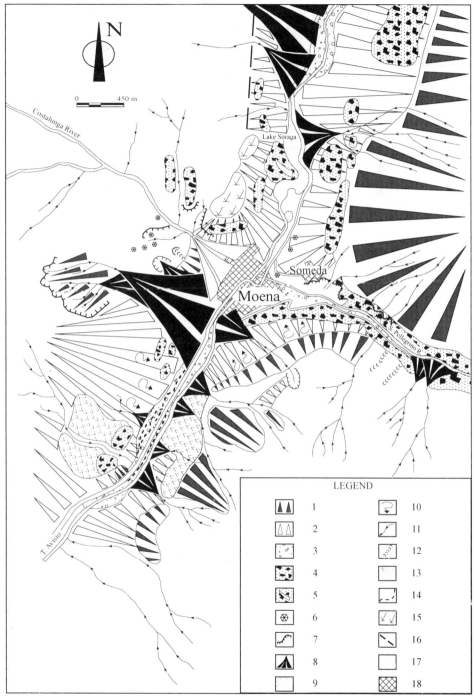

Figure 3. Geomorphological map of the area around Moena. Legend: 1) slope deposits; 2) mixed slope and fluvial deposits; 3) fluvial deposits; 4) glacial deposits; 5) maraine ridge; 6) interglacial conglomerate (large boulders) (Upper Pleistocene); 7) edge of degradation or landslide scarp less than 25 m; 8) mixed slope and alluvial fan; 9) area affected by small slope movements; 10) sheet erosion; 11) gully erosion; 12) small concave valley; 13) alluvial fan; 14) paleochannel; 15) marshy area; 16) fault; 17) bedrock; 18) built up area.

Slope deposits are particularly abundant and owe their origin to the gelidity effect on the bedrock. These deposits are usually the product of the accumulation of mixed processes, mainly caused by the actions of gravity and rillwash waters. In most cases they are characterised by the presence of silty and silty-clayey matrix; therefore they are often subject to sheet erosion and small slope movements such as soil slips and debris flows, too small to be represented on the geomorphological map.

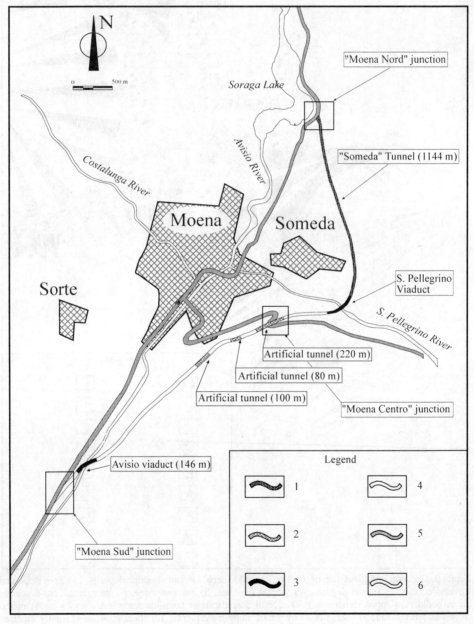

Figure 4. Illustration of the project. Legend: 1) tunnel; 2) artificial tunnel; 3) viaduct; 4) projected route; 5) present roads (state road 48 and 346); 6) hydrography.

In some sites, especially in the immediate surroundings of Moena, conglomeratic deposits consisting of mixed rounded clasts, seldom stratified and often covered by secondary carbonatic concretions, with silty-sandy matrix and size of 1 to 30 cm, often outcrop. The strong cementation and the high degree of weathering of many pebbles incorporated within the conglomerate have been interpreted as hints of maturity by Carton & Panizza (1983) who attribute to this deposit a pre-Würmian age, ascribable to an interglacial period after the Riss glaciation.

Today's landscape is the result of complex processes, some of which are no longer in equilibrium with the present morphoclimatic system. The slopes are in fact characterised by the presence of deposits and forms derived from the Last Glacial Maximum and the successive stadial episodes (Carton & Panizza 1983). Following the glaciers' withdrawal, glacial-type processes accomplished their activity and were progressively substituted by fluvial and gravitational processes. Slope denudation processes increased mainly beacuse of intense anthropogenic pressure on the area and caused the prevailing of gravitational processes on all the others. These are witnessed by several scree slopes, talus cones and fans. The high degree of activity of these processes is confirmed also by several examples of superficial slope movements which occur with one year frequency, mainly connected with extremely intense rainy events (i.e. in 1966, 1987, 1994, meteorological worsening). These processes are mainly related to two sets of causes: either intense summer storms or gelifluction phenomena in late winter - early spring period.

5 THE PROJECT

The project considers the construction of a new stretch of road by-passing the centre of Moena. The route is almost entirely developed on the left-hand side of the River Avisio for a total length of 3200 m, including the junctions (Fig. 4).

The starting point is on the straight stretch South of Moena, where the famous skiing competition "Marcialonga" is held, at km 45.5 of the Dolomite state road (ss 48) (Fig. 4). In this point a junction called "Moena sud" will be constructed. In the stretch between the "Moena sud" and "Moena centro" junctions (the latter will be at the intersection with the state road 346 of the San Pellegrino pass) the following works have been planned: a) a three-bay 146 m high viaduct with a height of 12 m above ground level at its highest point; it will permit the crossing of the Avisio River; b) some opencut stretches; c) artificial tunnels for a total length of 400 m (Fig. 4).

In the stretch between the "Moena centro" and "Moena nord" junctions, where the ring-road rejoins the state road 48, a 100 m high viaduct across the Rio San Pellegrino, directly connected with a 1144 m long tunnel (Someda tunnel), from altitude 1235 m to 1210 m, has been planned.

The volume of the excavated material following the digging of the tunnel was estimated at 150,000 m^3. The rock waste will be dumped South of Moena, outside the inhabited centre, near a horse riding centre, in an area belonging to the alluvial plain. The dumping site is placed on the left-hand side of the Avisio River and will occupy an area 15,000 m^2 wide and about 500 m long which will rise above the ground by about 11 m.

The preliminary environmental impact study, already carried out on behalf of the "Environmental Protection Service" (EIA office) of the Trento Province, has taken into account all the main components: geological attitude, atmospheric conditions, flora and fauna, landscape, historical and artistic elements, degree of development, production activities, farming and tourism. In this way, positive impacts (acoustic, touristic, economic aspects, pollution, transport times and conditions) and negative ones (geology and hydrogeology, landscape, tree cover) have emerged.

6 THE CASE STUDIED

6.1 *Geomorphological impacts*

For the evaluation of geomorphological assets and possible impacts on them, the methodology proposed by Panizza (1996) is used.

This methodology takes into account four characters (model of geomorphological evolution, object used for educational purposes, paleomorphological example, ecological support) in order to assess the intrinsic scientific value of the landforms and identify the preservation conditions and scientific quality of the landforms. The impact on landforms is therefore expressed in terms of reduction in scientific quality.

Table 1. Geomorphological assets of the area around Moena: polygon number (Poly); landform typology (Land) (10 = fluvial deposit, 11 = alluvial fan, 12 = paleoriver, ..., 71 = mixed fluvial and slope fan, ..., 80 = cemented deposit, ...); level of interest for the character: "model of geomorphological evolution" (Geom. evolution); level of interest for the character: "object used for educational purposes" (Educational purposes); level of interest for the character: "paleogeomorphological example" (Paleogeom. example); level of interest for the character: "ecological support" (Ecological support); geomorphological asset number (Asset).

Poly	Land	Geom. evolution	Educational purposes	Paleogeom. example	Ecological support	Asset
5	71	2	1	2	0	1
8	11	1	0	0	0	2
10	12	1	1	1	0	3
13	80	0	0	1	0	4
21	80	0	0	1	0	5
22	80	0	0	1	0	6
23	80	0	0	1	0	7
29	80	0	0	1	0	8
30	80	0	0	1	0	9
31	80	0	0	1	0	10
32	80	0	0	1	0	11
33	80	0	0	1	0	12
40	80	0	0	1	0	13
41	38	2	1	2	0	14

As a first step, the geomorphological assets were identified starting from the basic geomorphological map (Fig. 3). Afterwards the geomorphological map was subdivided into polygons which represent areas characterised by a particular morphogenetic type and nature (i.e. fluvial scarp). In this geomorphological map each polygon is linked to a table in which the following items are listed (Tab. 1):

1st column - progressive number of the polygon;

2nd column - code for morphogenetic process (first numeral) and landform (second numeral);

3rd-6th columns - level of interest (4 = worldwide; 3 = super-regional; 2 = regional; 1 = local; 0 = non significant) of each character (model of geomorphological evolution, object used for educational purposes, paleo-geomorphological example, ecological support);

7th column - number of the asset (the asset may be formed by more than one polygon).

Table 2. Impact on geomorphological assets of the area around Moena: geomorphological asset number (Asset); intrinsic scientific value (Value); condition of preservation (Preserv.); scientific quality (Quality); damage (Damage); impact (Impact). The total impact is 1.5.

Asset	Value	Preserv.	Quality	Damage	Impact
1	5	2	10	1	0
2	1	1	1	1	0
3	3	2	6	0.75	1.5
4	1	1	1	1	0
5	1	1	1	1	0
6	1	1	1	1	0
7	1	0.5	0.5	1	0
8	1	0.5	0.5	1	0
9	1	0.5	0.5	1	0
10	1	0.5	0.5	1	0
11	1	0.5	0.5	1	0
12	1	0.5	0.5	1	0
13	1	0.5	0.5	1	0
14	5	1	5	1	0

In this particular case, there are 14 selected assets in all: they are all simple, i.e. composed of a single form (Fig. 5). Among landforms the following should be mentioned: a large mixed fan (1) accumulated by means of gravity and rillwash waters, considered of regional importance since it gives paleogeomorphological evidence of morphoclimatic conditions different from the present ones; an alluvial fan (2) having only local importance since it is an example of geomorphological evolution; a system of abandoned fluvial courses (3) having local importance owing to its educational exemplarity and being also an example of geomorphological evolution as well as paleogeomorphological evidence; the polygons from (4) to (13) represent portions of the interglacial conglomerate. Polygon 4 represents a vast conglomerate outcrop whereas the others are only sporadic outcrops. They are all considered as locally significant since they make up geomorphological evidence. Finally, polygon (14) represents a frontal moraine ridge and glacial deposits witnessing a not better defined stadial phase and therefore assumes a regional value both as a paleogeomorphological evidence and as an example of geomorphological evolution.

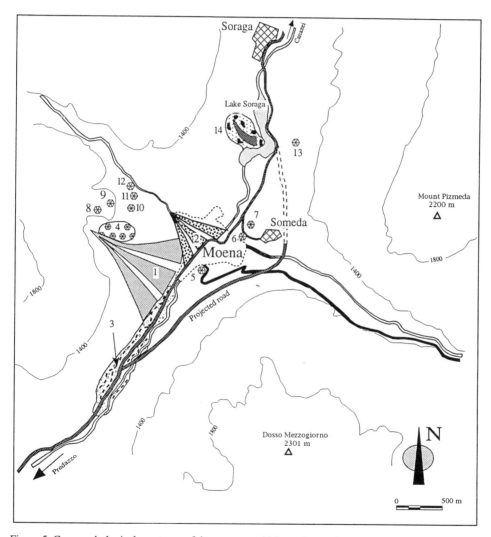

Figure 5. Geomorphological asset map of the area around Moena. Legend: 1) mixed slope and alluvial fan; 2) alluvial fan; 3) area occupied by paleoriver courses; 4) large interglacial conglomerate outcrop; 5-13) sporadic interglacial conglomerate outcrop; 14) moraine ridge and glacial deposits.

Table 1 can be subsequently linked to another table (Tab. 2) in which the following parameters are considered:

1 - the intrinsic scientific value of the landform (V) where V is the sum of each level of interest for all characters;
2 - the condition of preservation of the landform (C) ranging from 2, well preserved, to 0.5, poorly preserved;
3 - the scientific Quality (Q) of the landform where Q is the product V x C;
4 - the damage to the landform (D) ranging from 1, no degradation, to 0, complete destruction of the asset's characteristics;
5 - the impact (I) on the landform (I = Q x (1-D)).

Total assessment of the impacts can therefore be obtained as the summation of all the impacts on the landscape.

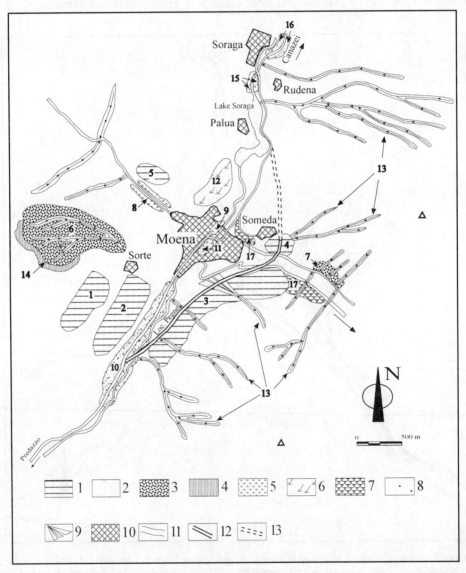

Figure 6. Geomorphological hazard map of the area around Moena. Legend: Areas subject to hazard due to: 1) surface gravitational movement and sheet erosion; 2) flood; 3) landslide; 4) scarp retrogradation; 5) bank fluvial erosion; 6) marshy area; 7) small slope movement; 8) gully erosion; 9) aggradding fan. Areas utilised as: 10) inhabited centre; 11) existing road; 12) projected road; 13) planned tunnel. The seventeen hazard areas are those of Table 3.

By observing the road route proposed, it is possible to notice that only in one case it does interfere with one of the selected polygons (Fig. 5).

The paleoriver traces could in fact undergo impacts, not only owing to the positioning of the access ramps to the "Moena sud" junction but also owing to the building yard activities during the construction phases.

The impact could be minimized by identifying adequate protection areas in the valley floor.

6.2 *Geomorphological hazards*

The qualitative geomorphological risk (in a broad sense) is expressed in terms of damage to the project or to the surrounding settlements.

In order to evaluate geomorphological hazard, in the study area, the first step was to prepare a geomorphological hazard map (Fig. 6).

In particular, some areas subject to flooding, surface gravitational movements, gully erosion, scarp retrogradation, marshy areas, landslides etc., have been identified in the geomorphological map.

A matrix for geomorphological hazard assessment was prepared (Tab. 3). In it the hazard types present in the area have been inserted in each polygon representing areas where geomorphological processes take place (Bollettinari 1995). For each of them the intensity is also evaluated.

Table 3. Geomorphological hazards on the project. Columns represent type of geomorphological hazard: river-bank erosion (ER); flood (ES); gully erosion (EL); scarp retrogradation (AS); area subjects to debris fall and accumulation (C); small slope movements (FR); gelifluction (G); washing out (RD). The numbers represent the degree of intensity: low or absent (0); medium (1); high (2).

	ER	ES	EL	AS	C	FR	G	RD
1	0	0	0	0	0	0	0	1
2	0	0	1	0	0	0	0	2
3	0	0	1	0	0	0	1	2
4	0	0	2	0	0	0	0	2
5	0	0	2	0	0	1	0	2
6	0	0	0	0	2	0	0	1
7	0	0	0	0	1	0	0	0
8	2	0	0	0	0	0	0	0
9	1	0	0	0	0	0	0	0
10	0	1	0	0	0	0	0	0
11	0	2	0	0	0	0	0	0
12	0	0	0	0	1	0	0	0
13	0	0	2	0	0	0	0	0
14	0	0	0	2	0	0	0	0
15	0	2	0	0	0	0	0	0
16	1	1	0	0	0	0	0	0
17	0	0	0	0	0	2	0	0

The processes deserving particular attention are those connected with surface slope movements. In fact, in these areas those processes are periodically developed which could be of some risk for the planned works. The remaining hazard types affect only marginally the work examined (floods, marshy areas, scarp retrogradation); anyway in these cases mitigations could easily be introduced (i.e. crossing over gullies in order to avoid linear erosion).

7 CONCLUSIONS

The use of simple methodologies (Panizza et al. 1996b) for the identification of geomorphological assets and hazards allows the role of the geomorphological component in Environmental Impact Assessment studies - which until recently had been neglected - to be better understood and assessed. In the study here considered the geomorphological assets identified are not affected by the project considered. From this point of view, therefore, it is not possible to recognise particular impacts on the landscape. As regards geomorphological processes, some situations which could result to be hazardous for the project itself are defined. These situations could be easily mitigated by introducing some construction devices which, by the way, have already been foreseen by the designer. From the geomorphological standpoint the implementation of the road project discussed is considered as not problematic for the environment.

REFERENCES

Bollettinari, G. 1995. Proposal of a method for qualitative and quantitative evaluation of assets, hazard and impact in the field of geomorphology. In M. Marchetti, M. Panizza, M. Soldati & D. Barani (eds), Geomorphology and Environmental Impact Assessment. Proceedings of the 1st and 2nd workshops of a "Human Capital and Mobility" project. *Quaderni di Geodinamica Alpina e Quaternaria* 3: 189-197.
Bosellini, A. 1996. *Geologia delle Dolomiti*. Bolzano: Athesia.
Bosellini, A., Castellarin, A., Rossi, P.L., Simboli, G. & Sommavilla E. 1977. Schema sedimentologico e stratigrafico per il Trias medio della Val di Fassa ed aree circostanti (Dolomiti Centrali). *Giorn. Geol.* 42: 83-108.
Bosellini, A., Neri, C. & Stefani, M. 1996. *Geologia delle Dolomiti. Introduzione Geologica. Guida alla Escursione Generale. 78 summer meeting, San Cassiano (Bz), 16-18 Sept. 1996*. Roma: Soc. Geol. It.
Carton, A. & Panizza, M. 1983. Geomorfologia dell'Alpe di Lusia tra Moena e il Monte Viezzena (Dolomiti). *Studi Trentini di Scienze Naturali* 60, Acta Geol.: 87-133.
Cavallin, A., Marchetti, M., Panizza, M. & Soldati, M. 1994. The role of Geomorphology in Environmental Impact Assessment. *Geomorphology*, 9: 143-153.
Marchetti, M., Panizza, M., Soldati, M. & Barani D. (eds) 1995. Geomorphology and Environmental Impact Assessment. Proceedings of the 1st and 2nd workshops of a "Human Capital and Mobility" project. *Quaderni di Geodinamica Alpina e Quaternaria* 3: 1-197.
Panizza, M. 1995. Introduction to a research methodology for Geomorphology and Environmental Impact Assessment. In M. Marchetti, M. Panizza, M. Soldati & D. Barani (eds), Geomorphology and Environmental Impact Assessment. Proceedings of the 1st and 2nd workshops of a "Human Capital and Mobility" project. *Quaderni di Geodinamica Alpina e Quaternaria* 3: 13-26.
Panizza, M. 1996. *Environmental Geomorphology*. Amsterdam: Elsevier.
Panizza, M., Fabbri, A.G., Marchetti, M. & Patrono, A. (eds) 1996a. Special EU Project Issue: Geomorphology and Environmental Impact Assessment. *ITC Journ.* 1995 4: 305-371.
Panizza, M., Marchetti, M. & Patrono, A. 1996b. A proposal for a simplified method for assessing impacts on landforms. In: M. Panizza, A.G. Fabbri, M. Marchetti & A. Patrono (eds), Special EU Project Issue: Geomorphology and Environmental Impact Assessment. *ITC Journ.* 1995 4: 324.

Basics of the natural river engineering

D. Komatina
Faculty of Civil Engineering, University of Belgrade, Yugoslavia

ABSTRACT: The impacts of traditional bank protection and river training works onto the river ecosystem are described in the paper. All considerations are based on the systematic approach - a river ecosystem is chosen as the system unit, which is decomposed into the physical, biological, and human subsystems. Interactions between the subsystems are discussed in detail. Many disturbing effects of the traditional works have caused a breakthrough of a modern approach - natural river engineering. Basic principles of the natural river engineering are presented in the paper, being closely related to the protection of natural channel patterns - pools, riffles, point bars, and bank vegetation. According to the principles, environmentally sound river regulation and maintenance procedures and techniques are described.

1 INTRODUCTION

The river engineering is concerned with the entire process of planning, design, construction and operation of works of various kinds, for the purpose of modifying natural river conditions to better serve human needs.

The river engineering generally involves the direct modification of rivers. The natural river environment and morphology is disturbed by each kind of work to a greater or less extent. Generally, the basic natural channel patterns are eliminated (pools, riffles, point bars, bank vegetation). The stream heterogeneity, which is the basic condition for preservation of aquatic and terrestrial organisms diversity, is thus reduced.

During the last few decades, as an environmental conscience was gradually being born, new approaches and techniques, known as the natural river engineering, were affirmated. All the procedures are based on the need to work with a river rather than against it, and to minimize the aesthetic and ecological degradation of a river and its environment. The natural river engineering measures mean not only a taking engineered rivers back into a stage, very close to its natural one (so-called river renaturalization), but also consider unengineered streams and give them a chance to be regulated according to the environmentally sound approach. The approach should be generally applied to small watersheds, since a degree of the ecosystem disturbance is much higher in such cases.

All those alterations of the landscape (or its parts) are regarded as "natural" when they are carried out with regard to the prevailing natural habitat conditions (the natural hydraulic and morphological ones). Non-natural alterations, on the other hand, can be regarded as those in which the existing natural conditions are not or are insufficiently taken into account (where only technical, mathematically simple solutions are sought). That is, beside customary conventional river engineering aims (as it is bank protection, or river improvement or training), there are now also ecological and aesthetic ones. Apparently, it requires a group effort based on comprehensive field work and background of ecological theory and experience.

Since the river engineering is one of the most dramatic aspects of human impact on the fluvial system, the impacts of traditional bank protection and river training works and the structures construction on aquatic and terrestrial animal and plant communities, are considered at the beginning of the paper. After that, the ecologically based techniques of bank protection works and river training (applicable to both already engineered rivers and unengineered streams), as well as the river maintenance procedures, are discussed.

2 ENVIRONMENTAL IMPACTS OF BANK PROTECTION WORKS AND RIVER TRAINING

2.1 *River as an ecosystem*

On the basis of a systematic approach methodology, a river basin is here chosen as the system unit. A river basin is a highly complex ecosystem which includes the organisms of a natural community together with their environment. In such ecosystem the interactions among various components are extremely complex. Hence, the system is decomposed into three subsystems (Boon et al. 1992): a) physical (comprising the sun energy, the atmosphere, water and earth, including soils and energy influences such as gravity and heat, and chemical factors as well), b) biological (composed of aquatic and terrestrial animals and plants including their interactions), and c) human (or socio-economic, covering all man's activities within the ecosystem), and then analyzed (Komatina 1996a).

2.1.1 *Biotic factors and processes in the ecosystem*

The energy cycle. Generally, there are three basic types of the ecosystem organisms directly involved in the energy transfer (Hynes 1970): producers (conversing the inorganic matter to the organic one); consumers (using the originated biomass) and reducers (disintegrating the organic matter to the inorganic one). The energy passes through these series and returns basic materials to their original form. The energy transmission is happened through the primary production (the rate of conversion of solar energy to the chemical one through photosynthesis; producers are all ecosystem plants), and the secondary production (the transmission led by invertebrates which, at the same time, represent primary consumers because they are feeding on the biomass originated through primary production). A food chain in the ecosystem is completed by secondary consumers which are feeding on the primary ones.

Succession. It can be defined most simply as a change in community composition after the ecosystem disturbance (Barnes & Minshall 1983). It involves two basic elements: colonization of the totally disturbed, neglected area (primary succession) and subsequent change of species due to the equilibrium violation (secondary succession). There are two succession types: site-specific temporal succession (communities change with time at the particular sites) and longitudinal one (sequence of communities in streams from headwaters to large rivers). According to small river basins analysis in the paper, a succession term will most often mean the first type mentioned.

Competition. A competition occurs when a number of animals (of the same or different species) utilize common resources the supply of which is short; or if the resources are not in short supply, it occurs when the organisms seeking that resource nevertheless harm each other in the process. Anyway, there is the joint utilization of limiting resources, and reduction in one or more components of an absolute organism fitness that results. There are two major competition types: exploitation (when competing individuals harm each other indirectly, by consuming and depleting a common, limiting resource) and interference (direct interaction between consumers, where some of them prevent others from gaining access to or using a resource).

Herbivory. The term means a consumption of living plants or their parts by animals. A primary effect on plant communities is removal of vegetation, which potentially reduces a biomass of primary producers. On the other hand, reducing standing crops in streams, herbivory is a dominant factor controlling the algae abundance.

Detritivory. Detritivory is the consumption of dead organic matter by animals.

Predation. Except under certain circumstances where competition or disturbance appear to play a critical role, predation is the most important factor controlling the plant and animal communities structure. The predation effects can be divided into two categories: those concerned with community-level events, and those concerned with evolutionary predator and prey responses to one another. The second one includes adaptations of the predator in choosing prey, and of the prey in minimizing its risk of being eaten. Generally, where predators are able to use vision to choose prey, a preference for larger prey appears universal. That is, fishes and other vertebrate predators show an increasing preference with increasing prey size. Invertebrate predators may use visual cues, but chemical, mechanical and tactile cues appear to be of a greater importance in detecting prey. Larger prey is preferred, but as prey size increases it becomes more difficult for the predator to subdue the prey. Virtually every group of aquatic insects includes at least some species that are predaceous on another invertebrates.

Coexistence. This means a life of different species with no predation-prey interactions. Usually, a microhabitat diversity may influence the diversity of species that coexist in the system.

Disturbance. Disturbance comprises all kinds of the ecosystem violation: fire and bank erosion, high discharges scouring and overturning stones in streams, ice formation anchored on a stone surface, scour due to drift ice (during spring ice break-up), and a human action. All of them violate the existing equilibrium in species interactions, thus perturbing the food chain to a less or greater extent.

Diversity of the ecosystem. The diversity means a number of species living there. The number can increase by: an increase in the range of resources available to and used by a community; a decrease in the range of resources used by an average species; and an increase in the average overlap in resource use between species. The diversity is a very important factor affecting the ecosystem stability.

Dominance. This is a term often considered and integrated with the system diversity and species relative abundance. The role of dominance may be an important aspect of community organization because the dominant species, those with high relative abundance, may so pervade certain ecosystems that they may effectively modify the ecology of other coexisting species.

Ecosystem stability. There are two stability concepts predominating in literature. The first is older one, based on the system variability. The systems which show a little temporal change (in population size and community composition) are considered as stable. Those with large temporal variability are considered as unstable. The concept supports a philosophy: the more system diversity, the more stability (Boon et al. 1992). The second one is recently developed and it is much better for human impacts assessment. The stability means the ability of an ecosystem to respond to disturbance. An ecosystem is considered stable if its response is small and its return to original state is relatively rapid. An ecosystem is unstable if it is greatly changed by disturbance, returns slowly to its original state, or if it never recovers to the original state. This concept comprises the two: relative stability and assimilative capacity. Relative stability has two aspects: ability of system to exhibit minimal response to disturbance and rapidity with which it responds. On the other hand, the assimilative capacity is defined as the ability of system to assimilate a substance without degrading or damaging the system ability to maintain its community structure and functional characteristics. The susceptibility (stability) and recovery time of various-scale habitats within the ecosystem is shown in Figure 1.

2.1.2 *A river ecosystem flora and fauna*

Algae. Algae grow attached to all kinds of solid objects, they also occur as thin films on mud and silt surfaces and, more rarely, as floating or attached in shallow littoral areas out of the current or among higher plants (Townsend 1980, Whitton 1984). There are three major types of algae: epipelic (living on mud), epilithic (living on stones or similar objects) and epiphytic (living on plants).

Flow velocity is an important factor affecting these organisms for various reasons, including the obvious mechanical one that the faster it is, the more attached species will be washed downstream. Very fast streams (above 5 m/s) reduce the flora to only those species which are both very firmly attached and also resistant to mechanical damage. The algal flora at such sites is usually dominated by encrusting forms and filaments with no or only moderate branching (*Chamaesiphon fuscus, Lemanea fluviatilis*). In slower streams the plant tends to form coralloid or plumose growths (*Stigeoclonium, Tribonema*).

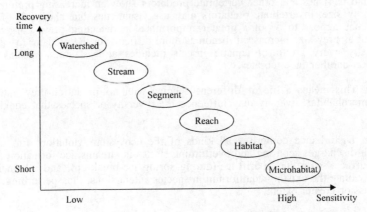

Figure 1. Relation between recovery time and system sensitivity to disturbance for different spatial scales.

Substratum may also influence the algae distribution. That is, the large and slower-growing species are often confined to the larger rocks (*Lemanea, Hildenbrandia*). In general, a coarse sandstone surface is particularly favourable for colonization, while a smooth surface is colonized more slowly. Limestones may be unfavourable for particular species or at least colonized slower than other rocks (Townsend 1980).

The algae distribution is also determined by temperature (Whitton 1975). As a mean temperature tends in general to increase from source to mouth, and temperature fluctuations are much less near the source than downstream, there is a logical species segregation at different sites (*Diatoma hiemale* prefers low temperatures; *Chrysonebula* prefers the higher ones). It is, however, often difficult to separate effects of temperature and light. Some algae are known to be very tolerant of shade (*Hildenbrandia rivularis*), some can survive living long periods in the dark (*Ulothrix*) and others need the high light intensities (*Cladophora glomerata*).

One of the most important chemical factors influencing the distribution of lotic species is the pH factor. It may influence growth of an organism directly (*Cyanidium caldarium*, with a pH optimum of about 2 - 3), and to be correlated with particular type of water chemistry.

Scour is very important feature of the algae ecology. During periods of active growth, particularly when the discharge is low and water is clear, the populations are built up and every solid object in the water may be covered with a thick algae layer. These populations are, however, very unstable and a single flood may wash them away. Much depends on the algae themselves and on the site where they are growing. Some species are little affected except by raging floods (*Rhodophyta*). Those growing on the stones rolled together by the water are much more affected (*Achnanthes*).

Grazing animals are also another factor. Many invertebrates and some fishes feed on some algae species (i.e. *Achnanthes*) and it is often possible to observe cleared stone surface caused by grazing.

Higher plants. Higher plants (macrophytes) comprise the remaining photosynthetic plants which occur on the water substratum. There are four major forms of it (Whitton 1975): emergents, rooted in the soil, but their leaves and reproductive organs are aerial (i.e. *Glyceria maxima, Phragmites communis*); floating-leaved, rooted in the soil with many of their leaves floating on the water surface and their reproductive organs are also floating or aerial (i.e. *Nuphar luteum, Potomogeton natans*); free-floating (i.e. *Ceratophyllum demersum, Eichhornia crassipes*) and submerged, attached to submerged material with their leaves entirely submerged, and with reproductive organs which may be aerial, floating or submerged (i.e. *Elodea canadensis, Fontinalis antipyretica*).

Flow velocity is one of the most important factors affecting the communities composition and location. Often the flows during floods are more important than the average flows (e.g. in a mountain stream *Myriophyllum spicatum* can be found in the fast flow zones, and *Elodea canadensis* in the slow ones).

The distribution of plant species is strongly correlated with the substratum nature. It is more important in a relatively slow, nutrient-poor water where fresh organic or inorganic deposits may be important sources of nutrients. If there is much material suspended in the water, the main effect will be through reduction of the light intensity, and slow growing species may be buried and eliminated by rapid silt accumulation.

Light is almost always an important limiting factor for plant growth. Although the upper leaves may be light-saturated, the ones within the plant stand, or deeper in the water, are receiving sub-saturating irradiance so that the whole community shows a response to the light level. In general, the submerged plants receive less light than emergent ones because of bank shading, surface reflection and attenuation by the water. That is one reason for their lower productivity. Under thicker tree cover the summer irradiance is reduced by some 95% and only few species of submerged plants survive. Plants can not grow in deep water, either because the substratum is unstable or because the water turbidity prevents light from reaching the bed.

The species found in a water of a particular chemical type are normally restricted to those that can grow well enough to succeed in competition. Few species are absolutely unable to grow in certain waters, but most can at least grow slowly in any water. Among chemical factors, it is difficult to isolate one as the causal.

Biotic factors are also to be mentioned. Submerged plants generally are not extensively grazed, but the emergents are often attacked by insects, birds and rodents. Further, once a new submerged plant of a species preferring high velocities becomes established, it spreads and then reduces the velocity around so that silt accumulates. These changed conditions may favour other species. For example, emergent species may become established, and the original colonizer is often eliminated.

Bankside plants. The plants may be divided into four major groups: trees, shrubs, grasses and herbs (Whitton 1984). They play vital roles in either providing shelter and protection for macroinvertebrates and fish, by shading watercourses from excessive water temperatures, providing an input of organic leaf litter, or as oxygen producers. Plants also diversify local conditions by altering the detailed flow patterns within a stream and by inducing silt deposition.

The most famous trees and shrubs types are: alder (*Alnus incana*), birch (*Betula pendula*), black poplar (*Populus nigra*), hybrid poplar (*Populus euramericana*), oak (*Quercus robur*), black locust (*Robinia pseudoacacia*), willow (*Salix purpurea*) and sallow (*Salix cinerea*). Often founded grasses and herbs species are: couch grass (*Elymus repens*), bellflower (*Campanula trachelium*), bindweed (*Convolvulus arvensis*), plantain (*Plantago lanceolata*), red clover (*Trifolium pratense*) and alfalfa (*Medicago sativa*). Factors affecting plants composition and location are the same as those mentioned earlier: climate, ground conditions, grazing, succession, etc. (Whitton 1984).

Microorganisms. Heterotrophic microorganisms are the part of the ecosystem decomposing organic matter to the inorganic one, thus completing the system energy cycle. Energy sources required by, and available to microorganisms in the ecosystem, are: autochthonous (plants and animals themselves, through death and decomposition and products of their metabolism) and allochthonous (soils, leaves and wind blown material). These are colonized by bacteria, fungi and protozoa and then decomposed. There is no doubt about the importance of heterotrophic microorganisms as decomposers but also as a group seriously affecting the oxygen regime of the system.

Macroinvertebrates. Those are primary consumers feeding on the biomass originated through photosynthesis (that is, the primary production) and also conversing it into a carbon dioxide. At the same time, these are secondary producers being eaten by higher, secondary consumers.

One of the affecting factors on these organisms is the flow velocity. Where the velocity is persistently high then almost all species may be absent. Conversely slow flows, which accumulate a large silt amount, can only be tolerated by a limited number of species. That is, in faster water species such as *Ancylus fluviatilis*, *Baetis carpathica* and *Rhithrogena semicolorata* can be found, and *Helicopsyche, Diura* and *Ameletus* occur in slower streams (Hynes 1970).

Another factor is the temperature. There are species living at high temperature sites (i.e. *Simulium latipes, Erpobdella monostriata*) and those that occur at low temperature ones (i.e. *Glossiphonia complanata, Helobdella stagnalis*).

Regarding the substratum, several species are adapted to live between stones, where turbulence is reduced. These species can only occur where the substratum is stony (i.e. *Ancylus*). On the other hand, some species require fine particles (*Goniobasis* lives on silt and sand, *Ephemera danica* on coarse sand). Riffles are typically characterized by a greater density of invertebrates than in a pool, although this may also depend on the presence of plants. In general the larger the stones, the more complex the substratum, and then the more diverse the invertebrate fauna. Species as *Baetis rhodani, Agapetus, Helodes* are usually abundant at more light sites, and *Tricorythus, Afronurus, Adenophlebia* are those living in shaded areas.

Taking dissolved oxygen into consideration, it can be seen that some species (i.e. *Piscicola geometra, Erpobdella octoculata*) require well-aerated water. On the other hand, *Glossiphonia complanata* and *Helobdella stagnalis* can survive in stagnant places.

Fishes. There are two major fishes groups (Amos 1970, Whitton 1975, Townsend 1980): reophilous (adapted to live in fast flows, with high dissolved oxygen concentrations) and limnophilous (which tolerate the relatively high temperatures and low dissolved oxygen concentrations). Salmonids (salmon *(Salmo salar)* and trout *(Salmo trutta)*) live in shallow streams with steep gradients and high flow velocity. In slower flows salmonids can be found together with fast-water cyprinids (barbel *(Barbus barbus)*). In the lower, slowest river stretches, the calm-water cyprinids (bream *(Abramis brama)*, tench *(Tinca tinca)* and carp *(Cyprinus carpio)*) are found.

Substratum consisting of rocks, boulder and pebbles is favourable for salmonids. On the other hand, the cyprinids live in a finer material (such as gravel).

Regarding a water temperature and dissolved oxygen concentration, the cyprinids (*Cyprinodon, Crenichthys*) are most often founded in warm streams with oxygen depleted, and salmonids prefer a well aerated and cool water. Most species have a wide tolerance of pH factor.

Competitive interaction between fish species is another factor seriously affecting their distribution. That is, some species are intolerant of competition from any others and tend always to occur more or less alone.

Relationships between fish species and other organisms are of a great importance for the ecosystem as well. Many mammals (otters and bears) eat fish, as do many birds. Fishes are also heavily parasitized by a great variety of protozoans and helminths.

Other vertebrates. These may be divided into four major groups: amphibians, reptiles, birds, and mammals. Some of them feed on invertebrates and fishes, others are vegetarian, but there are also very important predation and competition interactions, thus supplementing the system food chain.

2.2 *Environmental impacts of bank protection works and river training*

Stream channelization is an extreme physical ecosystem disturbance (Brookes 1988). Figure 2 compares the morphology and hydrology of a natural stream with a typical channelized watercourse. Channelization can alter the original channel dimensions and shape, the slope and the channel pattern, changing a heterogeneous system into a homogenous one (Komatina 1996a).

Bank cover is eliminated, pools are lost, flow approaches a laminar character, and the substratum approaches homogeneity throughout the channel. Boulders and debris may be removed to increase the channel hydraulic efficiency. Further, floodplain channelization reduces or eliminates the annual flood and the sediment load normally deposited in the environment. Levees prevent peak flows from flooding the adjacent land and therefore large quantities of nutrients and freshwater is lost. Both the abundance of water and nutrient supply are particularly responsible for maintaining the riparian ecosystem productivity and vitality. Channelization is frequently accompanied by extensive timber cutting so that the soil erosion and its consequences become a new problem to be solved. The result of the activities, in ecosystem terms, is that habitat diversity and niche potential are reduced and the

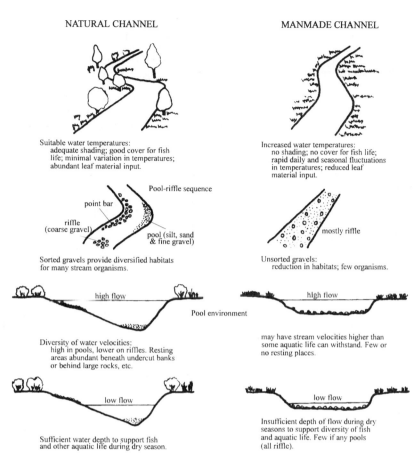

Figure 2. Comparison of the natural stream morphology with a channelized watercourse.

quality and functions of the species occupying the system are changed. How it affects the particular ecosystem populations further follows (Komatina 1992).

2.2.1 *Impacts on plants*
The destruction of the original substratum, significant changes of width or depth, and the emerged and bankside vegetation removal caused by bank protection and river training often have long-term effects on the vegetation. Many plant species are, simply, completely physically removed so that, depending on the material used for the structure construction, recovery appears to be an extremely slow process. Clearly, the largest ecosystem disturbance is the one caused by an use of concrete.

Dredging often reduces the morphologic variability of a channel bed and banks and this can reduce plant diversity. If the hard bed is unbroken by dredging, then plants recover quickly. If the bed is destroyed but silt accumulates, then plants may invade. However, if the substratum is unstable following works, and there is no associated sediment deposition, then the vegetation returns very slowly. Disrupting the bed removes underground plant parts, thereby delaying recovery. However, if dredging removes silt then the species composition may change.

Tall emerged plants recover relatively slowly after dredging because most have complex root systems which need to establish before they can dominate. Other plants, which initially regrow, are later shaded-out by tall species.

Over-deepened reaches within channelization works may function as permanent silt traps, thereby covering gravels and limiting the growth of submerged species. Deepening often replaces the original mixed sediment substratum with expanses of bedrock and the regrowth of vegetation is found to be inhibited. Boulders and gravels are removed from upland reaches, thus destroying the habitat for some species. Sufficiently deepened section may preclude the development of plants by reducing or eliminating the light available for plant growth at the bed. In over-deepened pools the accumulation of silt deposits may permanently obscure the substratum, thereby encouraging the growth of silt-loving plant varieties.

Weed cutting changes the percentage coverage of each plant species. Substantial amounts of silt can be washed out after the cut, and this enables some species to grow (i.e. *Ranunculus*). The maintenance in general suppresses the emergent species development.

2.2.2 *Impacts on macroinvertebrates*

During the structure construction process, macroinvertebrates may be physically removed or may fail to establish in the post-construction phase because of changed substratum. These changes are attributed to the loss of macrophytes as habitat, and to very unstable conditions in the substratum following the works, so that many susceptible species may disappear.

Silt deposition as a consequence of the works may kill many benthic invertebrates. Silt screens out light and tends to hold extraneous substances, such as industrial wastes, on the stream bed. Sand or shifting silt on the bed may also eliminate shelter.

Invertebrate drift is less in channelized reaches compared to natural ones, and this is partly attributable to the coarse substratum absence. The larger number of drift organisms is attributed to the lack of suitable attachment areas.

Channelization can result in decreased riffle habitat and gravel and boulder substratum. This is reflected in significant differences in macroinvertebrate density, species richness and composition, particularly a reduction in the abundance of few species (i.e. *Ephemeroptera, Trichoptera, Plecoptera*). It is attributed to differences in the substratum, pebbles being common in natural reaches and fine sand typifying channelized sections.

The timing of dredging may influence the degree of impact. A spring to summer dredge has the least effect, since most species breeding occurs shortly after redistribution.

The river narrowing by various structures reduces the water surface area and eliminates backwaters, thus decreasing habitat diversity. On the other hand, the structures create mosaic-like substratum - a habitat suitable for some species (i.e. *Hexagenia*).

The removal of snags from stream channels allows deposits of leaves, twigs and fine-grained sediments to be washed downstream. These deposits are an important habitat for many benthic species and in channels with sandy, shifting substratum form a suitable habitat.

Recovery of macroinvertebrates will depend on the frequency with which a channel is subsequently maintained. The annual weed cutting has a dramatic impact on the invertebrate fauna. The immediate effects are the removal of large numbers of animals in the weed, an increase in activity of many benthic species, and increased drifting of some plant-dwelling animals. The increased activity is attributed to the disturbance of the general habitat and to the supplementation of populations on the channel bed by animals escaping from the cut weed.

Recovery of the macroinvertebrate fauna following channelization generally occurs where there are no substantial changes in the substratum size and stability.

2.2.3 *Impacts on fish*

The effects of channelization are varied, depending on the nature of modification, intensity and extent, and subsequent morphological adjustment. Anyway, the effects are evident almost immediately after construction, whereas morphological adjustments take place in a longer term. It was also found that the average fish size in channelized streams is smaller than the one in natural streams.

The initial cause of fish populations moving out of a channelized reach may be a behavioural response. However, whether this is temporary or not will depend on the subsequent habitat changes caused by works.

Reduction of habitat diversity created by engineering works is most responsible for fish populations changes. A major reason for sparse populations in engineered stream reaches appears to be the lack of cover provided by undercut banks, overhanging vegetation, deep pools and other obstructions such as logs and boulders. A further effect of removing the bankside and stream plants cover may be to create excessive illumination and water temperatures. Rises in the temperature can not only affect the fish directly but can also eliminate invertebrates on which fish feed. Removal of bankside vegetation may also result in reduced invertebrate and fish populations as a result of the allochthonous (terrestrial) energy inputs loss.

Stream straightening may substantially reduce the available habitat. A commonly stated reason for the change is the loss of a natural pool-riffle sequence which provides a variety of flow conditions suitable as cover for fish and for organisms on which fish feed.

Shelter areas are required at high flows to protect fish from abnormally high water velocities and these conditions may be absent where a meandering stream, with an abundance of long pools, separated by short riffles is converted into a straight stream composed mainly of riffles.

Sediment released during dredging may have a detrimental effect on some fish species. Although fish can tolerate substantial amounts of suspended sediment, silt deposits can blanket portions of the stream bottom, reducing the available food by killing bottom-dwelling organisms.

As a consequence of the works, the river bed becomes uniform and unstable, together with an increase in suspended sediment concentrations and siltation. The mortality of trout eggs and fry is increased, thus causing a decrease in the fish populations numbers and biomass. Benthic invertebrate are declined and this could be correlated with the decline of fish which usually feed on these organisms.

Alteration of the width and depth variables in a channel may create shallow and unnatural flows, which result in an unsuitable habitat for fish and may present topographical difficulties for fish migration. Natural sorting of bedload materials on riffles and point bars is important in providing an environment in which bottom-dwelling organisms can thrive, thereby providing a food source for the fish.

The populations of species requiring sand and gravel substratum, rooted aquatic vegetation and well-defined pools and riffles and variety of flow conditions are either reduced or eliminated by the destruction of their habitats through channelization.

The time taken for recovery after channelization is an important aspect as well. The works type and extent appears to be an important factor, recovery being slowest where a channel is realigned and replaced by a straight channel. Projects which involve dredging, but retain the original watercourse, may recover more rapidly if pools and natural substratum materials reform.

Fish population may recover in the absence of maintenance. Weed cutting can have severe effects on river ecology and cause a decline of fisheries over several years. It effects fisheries by disrupting fish feeding and reproduction. Since aquatic macrophytes and algae form a major source of invertebrate fish food organisms, removal of these may be significant.

2.2.4 Impacts on other vertebrates

The severity of vegetational manipulation in the riparian habitat determines the degree of impact on the species composition.

The most sparse bird populations are found in areas which have been subject to vegetation removal, especially trees, shrubs and herbs, through bank protection and river training works. The degree to which various bird species are affected depends on their diet or procuring food method. Swallows, which are aerial insect catchers, and spotted pipers benefitted by the increased aerial space and gravel areas created by channelization. Conversely, hummingbirds are unaffected because animals on which they feed recover rapidly after the works. Species, feeding on insects from the stems and leaves of the shrub and tree, are most adversely affected by the engineering works. The recovery of bird numbers and species diversity is strongly corresponded with regrowth of the vegetation.

Bankside tree clearance, as a part of the channelization works, influences increasing pressures on the habitat, thereby causing continuing loss of, especially, small mammals species. The white-footed mouse population (*Peromyscus leucopus*), for example, recovers relatively rapidly because this species has a more general diet and habitat requirements and is able to exploit engineered areas. By comparison shrews (*Blarina brevicauda*) are non-existent for a much longer period, because they are more specific in their habitat requirements. Mink (*Mustela vison*), muskrat (*Ondatra zibethicus*) and beaver (*Castor canadensis*) are species much dependent on a riparian habitat as well.

2.2.5 General remarks

Man-made ecosystem disturbance creates a plenty of effects (Fig. 3), thus influencing a wide range of the animal and plant system communities (Boon et al. 1992, Komatina 1996a).

The vegetation removal caused by bank protection and river training appears to be the strongest shock for the system. That is, allochthonous organic particle input is reduced or eliminated with the elimination of bank cover, the channel is opened to direct solar input and essentials for autochthonous communities are developed. Production relative to respiration is increased, and structural diversity and the system stability are decreased. The vegetation destruction may extremely disturb also the macroinvertebrate fauna, according to a loss of macrophytes as a benthic invertebrate habitat. Fishery losses are attributed not only to vegetative cover destruction, but also to pools and riffles elimination, increased turbidity and reduced spawning areas, caused by the engineering works. Widespread destruction of natural plant communities may occur so that the habitats of birds and mammals may be destroyed by the trees, shrubs and grasses removal during construction, thereby affecting these vertebrate species.

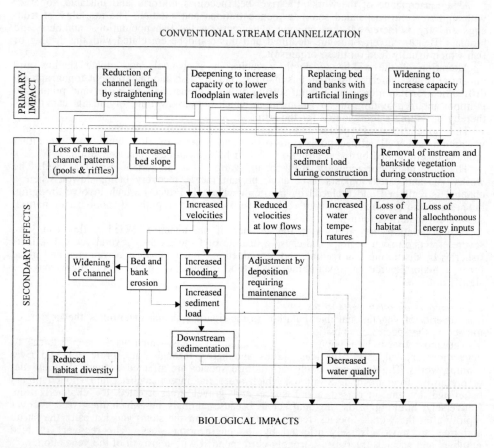

Figure 3. Channelization effects.

Bank protection and river training works also affect the downstream reaches. One of the most important effects is a downstream sedimentation. It simplifies the substratum and reduces species diversity, thereby decreasing productivity in the system across all its levels. Sediment can adversely affect fish eggs, particularly where deposition of silt is excessive. Deposits may blanket stream bed portions, thus eliminating potential spawning grounds and reducing the available food by killing bottom-dwelling organisms.

Downstream siltation due to erosion both during and after construction may destroy pools and riffles, and affect the fish populations. Suspended material may also smother the leaves, thereby preventing photosynthesis, or may be deposited above gravel causing plants to root in unstable conditions, or may lead to the burial of aquatic plant communities.

Time of recovery of the original sinuosity can vary from a few years, when the substratum is relatively fine-grained, to thousands of years when the substratum is hard bedrock, depending on the time necessary for a stream to become reinhabited by its original species.

3 NATURAL RIVER ENGINEERING

3.1 *Basic principles*

All the environmentally sound procedures rely on an understanding of the fluvial processes and forms of natural channels (Komatina 1996b). There are several fundamental components of the natural river channels that may be preserved or recreated to reduce environmental degradation in channel works (Fig. 4): a) Pool (P) - a topographically low area created by scour at high discharges. It is generally located immediately downstream from the axis of the bend and is characterized by relatively deep water and bed material usually composed of fine-grained sand. At the low flow the water is slow-moving; b) Riffle (Rf) - a topographically high area created by deposition at high discharges. Cross-section is typically symmetrical, the bed material is composed of larger rock sizes, often gravel, and the flow is fast at low discharges; c) Point bar (Pb) - the inner side of a bend is typically a area of deposition in contrast to the erosion of the opposite bank. Material accumulation forms a point bar adjacent to a pool, producing an asymmetrical profile; d) Floodplain - a natural river channel and its adjacent floodplain are parts of a single-system. Rivers occasionally overflow their banks and this process may be important in building a floodplain by deposition; e) Bank vegetation - it provides shade to the channel (preventing excessive illumination and water temperatures), bank stability and organic debris.

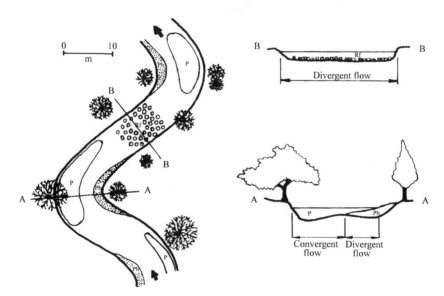

Figure 4. Idealized natural channel.

According to these components, few basic natural river engineering principles were created (Brookes 1988):

1. Channel plan should be based on natural forms, or those which have developed over long time periods. This also includes meanders and ox-bow lakes preservation, as well as the braided reaches in mountain streams.

2. Differentiation of the water course must either be preserved where it exists, or newly created where it is not found. It provides the necessary preconditions for a varied flora and fauna, and thereby a biologically intact water course.

3. Tree and shrub growth along streams plays a particularly important role in valley

landscapes which are often cleared of all vegetation. This may involve only a narrow vegetation belt or a more or less well preserved riparian woodland complex. The stabilizing function of the vegetation for the bank is as important as the shading of the water for regulating the oxygen content and limiting the growth of aquatic vegetation in nutrient-rich streams.

4. The preservation of the dynamics of meandering rivers and those transporting a bed load of rocky material is an objective which it is hard to fulfil in a densely built-up country. However, in rare cases it is possible to more or less leave rivers to their own devices. An availability of sufficient space is one of the most important preconditions for natural river engineering and unused space is becoming an increasingly rare commodity.

5. In the case of straight as well as meandering rivers, bank stabilization measures are unavoidable in order to limit the river dynamics, which might otherwise take up the whole valley. These measures can, depending on mechanical pressures placed on them, range from biological construction techniques through lightweight timber and stone structures, to the most heavy duty stone rivetting. They must all have one feature in common - the retention of the habitats of all aquatic and terrestrial organisms. This means for example that riprap banks must be designed in such a way both to provide habitats for animals which are fed on fish, and to provide shelter for the fish themselves.

6. The same principles apply to transverse structures. These should be designed in such a way that the aquatic fauna migration in both upstream and downstream direction is not impaired.

7. Regarding the embankments siting and structure, the safety must be the initial consideration, but it is also possible to create new damp meadow habitats by providing broad untouched areas in front of the structures. This objective is quite difficult to fulfil in built-up areas, but even here it is possible to create small natural stretches.

It is clear that these principles can only be met in reality if hydraulic engineers work closely together with biologists and/or other nature specialists. On this basis many different procedures and techniques have been developed. The procedures are considered in the next parts of the paper.

3.2 *Procedures and techniques*

3.2.1 *Biotechnical engineering*

Bank protection by vegetation. In several countries there has been an extensive research in the use of living vegetation rather than artificial materials for a bank protection (Petersen 1986, Komatina 1996b, c).

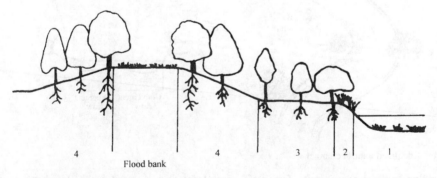

Figure 5. Zones of vegetation used for protection of a flood bank: 1) Aquatic plant zone; 2) Reed bank zone; 3) Softwood zone; 4) Hardwood zone.

Vegetation can be used to stabilize the channel bed, banks, floodplains, and embankments (Fig. 5). It performs a number of positive functions, as follows (Brookes 1988):
 1. interference (above-ground parts reduce flow velocity and hence erosion potential of flow adjacent to bank);
 2. protection (above-ground parts shield the surface soil particles from erosion by the flow);

3. restraint (surface root structure restrains entrainment of surface soil particles by flow);
4. reinforcement (underground root structure reinforces the subsoil at depth);
5. anchorage (underground root structure anchors the surface vegetation into the subsoil);
6. buttressing (individual plants restrain a wider soil mass from movement);
7. transpiration (soil moisture drawn in by the roots is released through the foliage; in the extreme, this can cause negative pore pressures in soil).

When planning bank protection works by using vegetation, it is important to know that the soil type, soil water regime and climate are key factors determining the ability of a site to sustain vegetation.

Figure 6 illustrates four distinct vegetative bank zones. A summary follows of the role in bank protection of the plants found in each zone (Coppin & Richards 1990, Komatina 1996b).

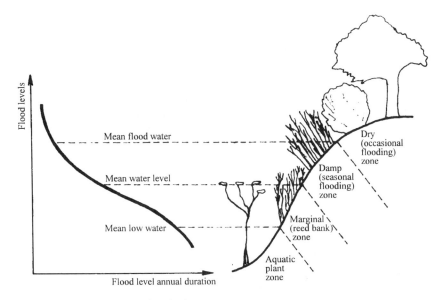

Figure 6. Vegetation zones on a river bank.

1. Aquatic plant zone. Plants in this zone will survive only with a slow current and sufficient light. They produce minor protection, but significant potential to reduce the channel capacity, so that considerable effort is expended on their clearance.

2. Marginal (reed bank) zone. Marginal plants can form an effective interference zone for restricting flow velocity, as well as providing protection and restraint. Root reinforcement of surface soil forms a protective soil-root mat. They also enable the sediment accumulation by the dense plant stems.

Unless flow velocities are very low, reeds are weakened by the loosening of soil around the root structure and it is normally necessary to protect their roots (if their roots are protected, reeds are able to stand up to quite high velocities). In faster-flowing waters, plants themselves may be ineffective and it is often necessary to protect critical areas using inert materials. Riprap is commonly used and there are many techniques utilizing geotextile meshes and pocket fabrics.

Commonly used reeds types are bulrush (*Schoenoplectus lacustris*) and sedges (*Carex*), but the latter can be invasive and reduce effective bed width on smaller streams.

3. Damp (seasonal flooding) zone. This is a natural habitat for fast-growing shrubs and trees, such as willow and alder, whose roots tolerate submergence. Vegetation effects are the same as for marginal zone. Trees do not give protection but roots provide a buttressing effect.

Trees should be planted in groups, not in a continuous line, so that provide access between them for future maintenance. Alder (*Alnus incana*) should be planted some distance back from the bank top to form a sacrificial area which can be allowed to erode until tree roots are sufficiently established to hold the bank (Fig. 7). Alder strikes strong vertical roots and thus when established is the most useful protection. It can, however, grow very high and should be cut to prevent the development of large overhanging trees which may eventually fall and destabilize the bank section (Brandon 1989).

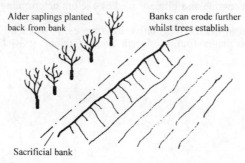

Figure 7. Planting of alder.

Poles cut from either crack willow (*Salix fragilis*) or white willow (*Salix alba*) may be used to repair a washout or hole in an existing bank or protect a new bank (Fig. 8). Alternatively, willow poles are packed into the front and top of gabion baskets (Fig. 8).

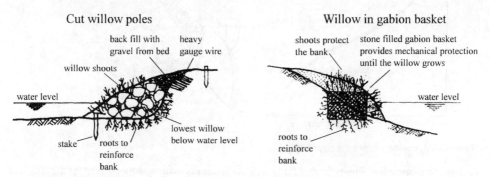

Figure 8. Use of cut willow poles.

Osiers (*Salix viminalis*) are used in faggots and in spiling. Osier sets can be pushed into a gravel or sand bank following dredging, where they will soon root and produce abundant and bushy growth.

Willow and alder, combined with a geotextile as a composite material, are frequently used to give both immediate and long-term protection. It can also be planted in the joints of riprap to bind the structure together.

Grass can only be used above the normal summer water level because its roots do not survive prolonged submergence. Where very high flows are expected, geotextile-reinforced grass may be used.

4. Dry (occasional flooding) zone. Trees are less significant in bank protection in this zone because it is less often flooded and less easily eroded by water. They are useful means of controlling by shading the aquatic plants growth (Fig. 9). The shade amount has to be sufficient to control aquatic vegetation, but not enough to shade out the shrub, grass or reed. Grass is also often used to protect the dry zone and this must receive sufficient light to grow and function properly.

Grass as an engineering material is very efficient and widely used. If the sward is dense and well managed, grass will withstand considerable velocities. To give added protection under severe loading conditions, grass may be reinforced by the addition of a geotextile fabric, net or mat. These geotextiles are either pinned in place on a soiled and seeded bank and adequately overlapped, or laid under a thickness of topsoil and seed. In the latter case, which is preferable for fabrics and mats, the roots grow down through the geotextile, forming a complete and integrated armour layer (Coppin & Richards 1990).

In rivers with braided flow or very high seasonal level fluctuations, trees or shrubby vegetation can be used for river training. Planting is mainly in the form of retards, groynes, or close-planted belts usually located on the edge of peak flood flow.

Orientation	Bank for shade belt	Effectiveness for weed control
E — W	South bank	Maximum
SE — NW	South-east bank	Moderate to good
SE — NW	South-west bank	Moderate to good
N — S	East bank East and west bank	Poor Moderate

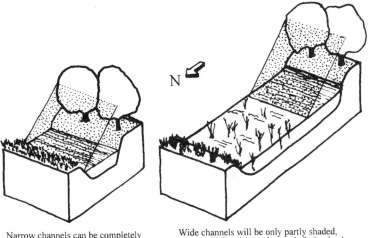

Narrow channels can be completely shaded from one side

Wide channels will be only partly shaded, plants still flourish on the south-facing bank

Figure 9. Tree planting for shade.

Bank protection by other natural materials. The possibilities are the following (Komatina 1996c):

1. Natural native hardwoods. Those, such as oak, ash and elm, will afford physical protection for perhaps 5 to 10 years, if built into the bank bottom in the intermittently wetted zone. Out of the water they can last longer. Often used is thorn, particularly where the soils are generally weak and can not accept any appreciable surcharge loading (Brandon 1989). The thorn faggots (Fig. 10) both reinforce weak soils and provide drainage.

2. Wattle hurdles. They are normally woven from hazel saplings into protective panels. They are used horizontally to protect a newly formed bank from erosion until adequate vegetation is grown to give permanent protection. Laid on the bank slope, they have a limited life but are cheap, effective and environmentally acceptable. Being biodegradable, they are more satisfactory from the long-term environmental standpoint than their more modern geotextile equivalents.

3. Hardwood timber piles and toe boarding. Hardwood boards are usually used to support the bank toe and protect up to the average winter flow level.

4. Faggoting and spiling. Both these bank protection methods use a combination of brushwood and stakes to reinforce the lower slopes of erodible banks. If faggoting is executed with dead brush, the natural sediment and organic matter accumulation creates an

important habitat below water level for invertebrates. Above the water level this consolidated bank enables plant communities to become re-established. Spiling or faggoting with willow creates a totally different environment. The matted roots which project into the water are rich in invertebrates, and are important egg-laying and fry habitats for fish. The food and cover provided by the shoots enable many insects, birds and mammals to utilize the river bank.

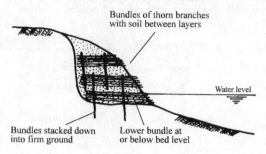

Figure 10. Thorn faggots.

3.2.2 *Specific areas alternatives*

Two-stage channels design. The two-stage channels concept is often used for upstream urban areas alleviating flooding. These channels confine the normal flow range to the original channel, while flood-flow is contained within a larger channel constructed above bankfull by widening out the floodplain (Fig. 11). Low-flow channels prevent excessive sediment deposition in the enlarged channel and enable the fish migration and avoid excessive water temperatures associated with conventional flood channels (Brookes 1988). A wetland flora may develop on the berm areas.

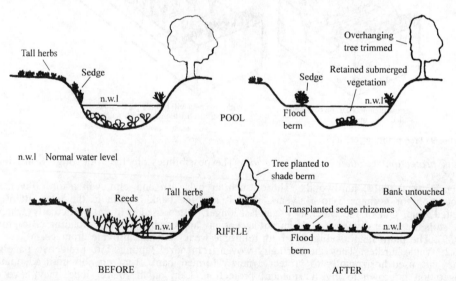

Figure 11. Two-stage channel design.

When constructing a new low-flow channel it is desirable to emulate the morphological characteristics of the original channel, thereby providing stability, and biological and aesthetic diversity. When the existing natural channel is retained as a low-flow channel the original substratum, bedforms and alignment are preserved. If the original stream has sufficient power to erode its channel then it may be necessary to restrain the low-flow channel from migration.

Partial restoration. This procedure was developed for urban streams restoring. It reproduces some natural streams characteristics, but at the same time achieves the engineering flood prevention objective. Stream restoration involves removal of urban trash, extensive growth of small trees and bush, and removal of large trees fallen into the stream channel (Fig. 12). The process does not involve straightening or the bed enlarging or removal of all trees along the banks. Trees are left intact, the root system helping to control erosion and increase aesthetic quality (Brookes 1988). The inside slope of each bend is angled at 1:3 or less. This promotes the point bar development on the inside of the bend as found in natural channels. The outside of bends may be ripraped where absolute bank stability is required. Alternatively, the bank can be planted with grass. This procedure trades a small loss of flow efficiency for a more stable and biologically productive morphology.

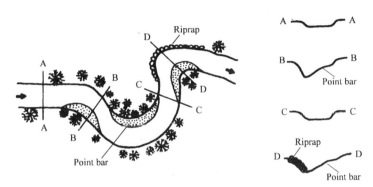

Figure 12. The channel restoration concept.

Cut-off meanders. In many classical river improvement schemes the meanders loss was an almost universal feature of the work. In the straightening process not only the cliff habitats were lost, but problems also were found downstream as a result of shortening the watercourse. Both on the eroding meander cliff face and at the shallow shelves on the inside of the bend, their characteristic wetland plants and animals decline (Brandon 1989).

Figure 13 illustrates three methods of retaining a meander by a cutoff channel which creates greater capacity, reduces the erosion impacts, and does not effectively shorten the river length. However, whatever the variation tried, the cliff and shelf habitat should remain, flow through the old channel must continue during floods, the old channel must contain water at all times, ponded or flowing, and the new channel is so aligned as not to cause erosion downstream.

Option A (Fig. 13) is the cheapest, but offer the least enhancement. It is ideally suited to a grazing land. Examples B and C create major changes to the old channel but retain the meander habitats intact. The choice of option will be determined by many factors, including cost, individual site and the substratum characteristics.

3.2.3 *Floodplain approaches*
This category includes those procedures which utilize the floodplain for flood control as an alternative for modifying the existing channel. Such approaches enable existing morphologic and biologic characteristics to be preserved (Komatina 1996b).

Corridors. For rivers which are actively changing their courses, it is suggested that management must provide a sufficient corridor to allow continued change (Fig. 14). Although a river might shift over the entire floodplain in a period of several hundred years, migration of individual bends is usually accomplished in a few tens of years. Thus, the meander belt may be confined to a fixed position on a floodplain over a period of time for planning and management. Geomorphologists can attempt to determine the corridor width and location that can be maintained for planning objectives (50 to 100 years). Although river response cannot be accurately predicted, the corridor concept may provide a valid alternative to confining a channel to a single position by either straightening and/or extensive bank protection (Brookes 1988).

Within belt meanders, braided reaches and pools and riffles are preserved. This maintains natural habitat conditions. The preservation of old meander loops and land adjacent to the channel provide habitats for a variety of flora and fauna. There are some recommendations for the watercourses care:
1. acquire abandoned meander loops and connect to provide water flow;
2. acquire stands of trees on the former meander loop;
3. preserve stands of polars and other indigenous species;
4. acquire reed bank in the former meander loop;
5. acquire land with the remnants of the water meadow forest trees growing on the floodplain;
6. plant the river banks with indigenous species, particularly to protect the outer bank;
7. maintain existing trees growing along the watercourse.

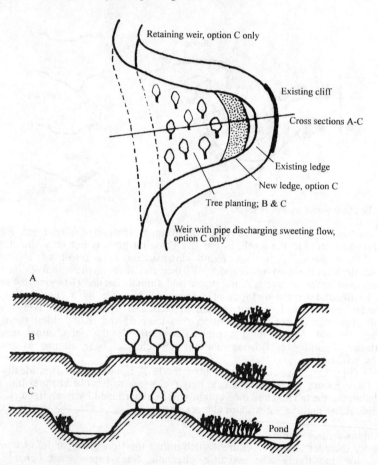

Figure 13. Retaining meander by cut-off channel.

For artificially straightened channels in Denmark which are regaining their former sinuous courses in the absence of maintenance, it has been proposed that a corridor be delineated within which the stream continue to migrate through erosion and deposition processes. This can be a far more economic solution than using extensive engineering methods to constrain a streamcourse. Corridors will preserve instream habitat and avoid the erratic responses associated with more conventional channelization procedures. Clearly, a corridor cannot be created in those circumstances where continued channel migration threatens property.

Floodways. Floodways or bypass channels are separate channels constructed on floodplains into which high flows are diverted. The existing natural channel is left unaffected.

3.3 River renaturalization

3.3.1 Revised construction procedures

The adverse effects of conventional methods may be minimized by careful options selection at the planning stage, or by a limiting degree to which a channel is modified (Komatina 1992, 1996b).

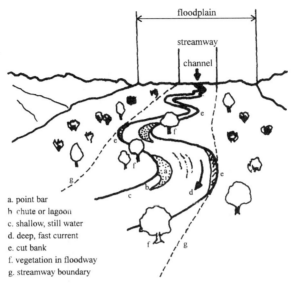

a. point bar
b. chute or lagoon
c. shallow, still water
d. deep, fast current
e. cut bank
f. vegetation in floodway
g. streamway boundary

Figure 14. The streamway concept.

Realignment. At the design stage, there should be minimal channel length reduction, the excavation and fill amount should be controlled, and equipment which minimizes bank and streamside destruction should be used. Banks should be replanted wherever possible and riprap placed such that the vegetation growth near the stream edge is not impeded. During construction, access by vehicles should be strictly controlled. Minimal streambed and bank disruption can be attained by educating foremen and specifying the types of equipment that can be used in particular areas. Finally, in clean-up phase, it is recommended that gravel and larger rocks are replaced in the bed to approximate conditions existing prior to construction and to restore stability. Banks replanting and reseeding with native trees, plants or grass provides shelter and cover for wildlife.

Meander preservation. Where the channel location must be moved, meandering alignments may be more expensive to construct than straight channels because of increased excavation costs. However, environmental benefits and reduced maintenance costs may offset the increased construction costs over the project life. Well-designed meandering channels are more stable, provide a greater variety of flow conditions and aquatic habitat diversity, and are aesthetically more pleasing. To design a suitable alignment, it is recommended that the existing meandering geometry and slope be used as a guide, based on field surveys, maps, or aerial photographs. The size, shape, meander geometry, slope and bed roughness should be similar to the old channel.

Channel enlargement. Channel enlargement by modifying only one bank, leaving the opposite bank almost entirely untouched, is now a common practice in many countries. Vegetation on the opposite bank is disturbed as little as possible, although potential obstructions to flow such as individual trees may be removed. The bank from which the work is undertaken can be designated on the basis of habitat value of the vegetation, aesthetics, shade and bank stability. If work is alternated from one bank to the other, the aesthetic appearance may be improved and sensitive habitats can be avoided. Retention of tall vegetation will shade the aquatic one, thereby reducing maintenance costs. If the channel is widened, then all vegetation on the working bank will be lost. However, where widening is not significant, it is beneficial to retain as much vegetation as practical. Damage can be minimized by using small equipment and by revegetating disturbed areas.

The excavation impact on the aquatic vegetation is minimized by avoiding the creation of very deep pools which may serve as silt traps or preclude light from reaching the channel bed. It is also recommended that excessive widening be avoided because this is likely to reduce the water depth (for a given discharge), thus limiting vegetation growth. Pools and riffles should be preserved wherever possible.

Embankments. Since typically embankments have an uniform and monotonous appearance, they often lack ecological, recreational and aesthetical values. Therefore, it is useful to know few guidelines for environmentally sensitive flood bank construction: flood bank should be set back from the immediate channel edge, especially if the river bank is to be protected; it may be desirable in other circumstances to create a shallow wet marginal zone for reeds; further, trees and shrubs retaining on the inside of a new flood bank, and their planting between the river and embankment, should be considered in order to enhance habitats for wildlife and visual appearance; where annual maintenance is required use should be made of herbs and slow-growing grasses, and trees and shrubs would be inappropriate.

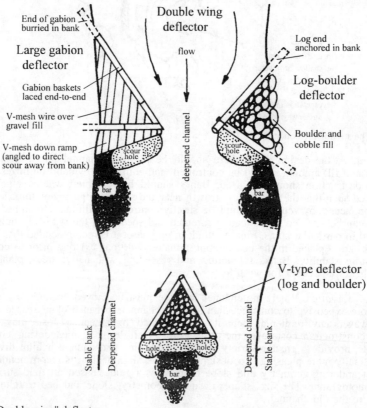

Figure 15. "Double wing" deflector.

Lined channels. The channels lining and paving creates a very unnatural bed and usually accompanies channel enlargement and/or straightening. Measures may be taken to minimize the adverse impacts including the placement suitable bed materials and vegetation retention or planting. A curved alignment could be used.

The choice of materials for bank protection is the most important from the environmental viewpoint. Rigid linings such as reinforced concrete, grouted riprap, bagged cement, and filled mats have perhaps the most detrimental effect on the aquatic habitat. By contrast, riprap of cobble or rubble, gabions, gravel armouring, grasses and woody vegetation are much more desirable. Although these linings do not provide a habitat for the same benthic community as the natural channel, they are much better than concrete and can be just an effective in erosion preventing. Vegetation can be used alone or in conjunction with structural protection, but it is imperative that it becomes established before the next flood.

Experience with enhancement techniques has revealed the advantages of riprap, composed of natural rock or quarry stone, as opposed to more conventional protection methods. Vegetation can be established between the stones, thus providing a stable substratum for benthic invertebrates and the weathering of stones may produce gravel bed suitable for fish spawning. Increased primary productivity and invertebrates abundance is associated with riprap as well.

Concrete channels may be alternated with short lengths of natural channel which provide an acceptable habitat for fish. Anyway, the length of bank protection measures should be limited to the absolute minimum required for stabilization.

Dike fields. While the major objective of dikes is to stabilize relatively long lengths of river, they can also provide an extremely valuable habitat for fish and macroinvertebrates. The main problem is to design dike fields which do not fill with sediment. This can be achieved by varying the length and height, but constriction gaps or notches in dikes are the most widely used environmental feature at present. These allow water to flow through the dike at intermediate stages and prevent sediment accretion by scouring. A variety of notch widths, shapes and depths are recommended through a reach to provide spatial and temporal habitat diversity.

3.3.2 *Mitigation and enhancement methods*
It is possible to mitigate or enhance a channel against the adverse effects of channelization by installing structures or taking measures at some time after construction. These measures can be used in the planning, design, construction or maintenance and they improve the net environmental effect (Komatina 1992, 1996b).

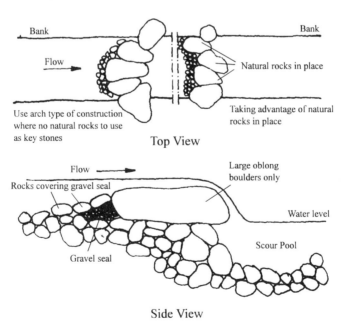

Figure 16. Boulder dam.

Instream devices. Deflectors function to either direct flow and eliminate accumulated sediments or to narrow a channel, thereby increasing velocity and creating a scour pool with a corresponding riffle downstream. A spacing of five to seven channel widths is recommended, corresponding to the pool-riffle spacing found in natural streams (Gore 1985). The basic deflector types are rock-boulder, log-boulder, double wing deflector (Fig. 15), and gabion deflector.

Low dams and weirs are partial barriers to flow which extend across the full channel width. Weirs generally do not extend above the water surface and serve to break up river flow and increase turbulence. The impoundment low dam effect increases upstream water depth and reduces flow velocity. The overspill from the dam both aerates the water and produces a scour pool below the dam (Gore 1985). A shallow, fast-flowing riffle area is then created where the excavated bed material is deposited further downstream. Low dams are probably among the most effective devices for producing a pool-riffle pattern in small rivers (Hara 1993). The most often found are boulder dams (Fig. 16), log dam (Fig. 17) and gabion check dam.

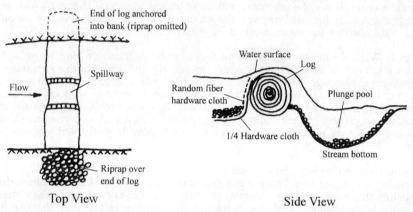

Figure 17. Log dam.

Substratum development. The riverbed form is of major importance to aquatic communities, and both benthic invertebrates and fish can be severely affected by disturbances which modify the size composition and reduce the bed material stability. Various instream devices can be used in a variety of ways to improve substratum conditions. Both current deflectors and low dams can scour deposited sediments by increasing flow velocity. Low dams and weirs can be also used to increase substratum stability in high-gradient streams. Gravel and other suitable substrates can be reintroduced to the streambed to restore lost spawning areas for salmonids.

Pools and riffles. The procedure for the pools and riffles design depends on whether they are to be constructed in lined or unlined channels, whether additional material is to be added to the channel, or whether the bedforms are to be allowed to reform naturally in absence of maintenance. As a preliminary step, it is necessary to assess the flow characteristics and channel morphology to determine if pools and riffles are appropriate instream habitat features. The main criterion is usually the stream ability to support a fishing resource. Pools and riffles are not usually installed in steep-gradient channels where there is a high sediment transport, or where the banks are unstable. They are rarely constructed where the bed material is too coarse to be moved under present hydrological conditions.

The pools and riffles spacing is not critical in lined channels. For unlined channels, spacing can be determined from neighbouring streams with similar characteristics or from other reaches of the same watercourse. Generally, an average of five to seven channel widths are sufficient to emulate natural conditions. Regular spacing should be avoided. A meandering alignment should be incorporated, the riffles being located in straight reaches and pools at the bends. Proper spacing facilitates self-maintenance. Pools and riffles dimensions may be varied to suit habitat requirements.

In cases where the riffles are to be dynamic and self-maintaining, they should be constructed from natural stream gravels with a size distribution typical of the existing bed material. Otherwise, they can be constructed from gabions, cobbles, or riprap.

Other structures and treatments. Instream and bankside cover is one of the most important of all habitat features for fish populations. Submerged cover (vegetation, rock and boulders, organic debris, etc.) has many roles in the river environment. Fish utilize cover for shade, shelter from predators, and a spawning substratum, while invertebrates associated with aquatic plants and debris provide an abundant food source.

Bankside cover provided by overhanging trees and bushes is particularly important to fish for shade and cover. Submerged tree roots, branches, and undercut banks also provide fish with shelter, food, and spawning areas.

Most artificial cover devices are designed to simulate bankside cover areas. These structures usually take the form of overhanging platforms extending along river margins, either floating on the water surface or supported by pilings driven into the riverbed. Bankside cover can also be provided by anchoring felled trees and large branches to the bank and allowing the foliage to trail along the river margins. Rocks and boulders installed along a riverbank to reduce erosion can also provide valuable cover for small fish.

Instream cover devices can also be installed in the main river channel to provide direct cover and to increase flow diversity and water turbulence. Large boulders arranged along the riverbed are one of the most widely used forms in gravel-bed streams (Gore 1985). The main objectives of boulder placement (Fig. 18) include the provision of additional habitat and fish cover, improving and restoring pool-riffle and meandering patterns, and increasing flow diversity (Hara 1993). Trash catchers are easily constructed devices used to create pools, provide cover, slow velocities, and hold spawning gravel in place.

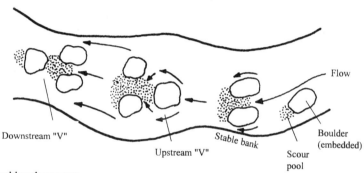

Figure 18. Boulder placements.

3.3.3 *Reconversion*

This category includes those instances where a channelized reach is reconverted to the original natural channel. It can be undertaken only where an enlarged or modified channel is no more required to meet an engineering objective (Komatina 1996b). Few procedures illustrated by practical examples will be described.

Sizing channels. Aggraded channels in North Carolina (USA) have been restored to their former capacities by sizing neighbouring natural channels. Manning's equation is used to determine the discharge of the natural channel cross-sections, which are then related to the drainage area. The aggraded channel capacity is then determined and compared to that of idealized natural channel for the same drainage area. The excavation degree can then be estimated.

Restoring the sinuosity. The former sinuosity, cross-sectional dimensions, slope and substratum of a small, straightened stream have been recreated in southern Denmark. The restored reach morphology was determined from historical maps, from comparison of naturally sinuous streams in neighbouring catchments with similar physical characteristics. The recreated course required stabilization by riprap on the outside of the bends before vegetation became established. Gravels were placed on the stream bed to recreate the substratum. The new course replaced a severely degrading straightened reach and, because of the restored morphology and hydrologic diversity, the channel has been successfully colonized by a variety of flora and fauna.

Roughness elements. In streams in northern California (USA) the habitat for salmonids has been improved by reintroducing large-scale roughness elements. These include boulders and woody debris which change the distribution of hydraulic forces over a streambed and thereby cause scour and the sorting of fine sediment from gravel.

3.4 *Alternative maintenance procedures*

Environmental impacts may be considerably reduced by partially maintaining channels. Recovery of channels following major works may occur in the absence of maintenance, while conventional maintenance practices, which disrupt the channel bed and banks, may be substituted by less detrimental techniques such as selective clearing and snagging, dredging, and shading (Komatina 1996b, d).

3.4.1 *Biological recovery*

Since maintenance usually destroys the habitats of various organisms by removing the bankside vegetation, encouraging bank instability, and preventing the development of a stable substratum, biological recovery from conventional channelization practices may occur in the absence of maintenance. Recovery is dependent on an improvement in habitat, but without intervention of man it may be a very slow process (taking between 50 and 100 years).

3.4.2 *Selective clearing and snagging*

Clearing and snagging may be defined as the removal of woody vegetation and debris from stream channels and banks to preserve flood capacity and minimize erosion. Traditionally, all vegetation has been removed to the detriment of the environment. Therefore, a number of alternative procedures have been developed. These alternatives focus on limiting the amount of vegetation which is removed or using techniques which minimize disruption to the environment.

A general guidance is that clearing and snagging should only occur at locations where significant blockages occur. Logs should be removed only if they obstruct flow, causing upstream ponding and/or sediment deposition, or if they are free. Logs which are rooted in the channel or floodplain and not causing blockage problems should not be removed. Rooted trees should be removed only if they are leaning over the channel at an angle of 30° or more or are dead and likely to fall into the channel within one year and create a blockage to flow. Very small accumulations of debris or sediment need not to be removed unless they obstruct flow to significant degree.

Hand removal or hand-operated equipment is preferred for the removal of logs. Alternatively, a small water-based crane or small crawler-tractor with winch may be used. Access routes should be carefully selected to avoid damage to floodplain vegetation. Rooted trees should be cut well above the base, leaving roots and stumps undisturbed. Cut logs should be cleared away from the floodplain to avoid the risk of their reentering the channel through over-bank flows.

3.4.3 *Dredging*

Although channel dredging has the greatest potential to damage a river ecosystem, it is the river maintenance activity which also offers the most enhancement potential.

The following are examples of procedures which ensure that the procedure, at worst, retains existing features of a river wildlife, but also offers great potential to improve habitats for plants and animals.

Retaining or creating riffles. Figure 19a shows the ideal approach when attempting to retain riffles created by gravel deposition or hard bed material. The longitudinal section shows that the ideal approach is to excavate a relatively uniform amount of material from the whole length being dredged. To retain the habitat diversity, a similar depth of material should be removed from the pools and the riffles (Brandon 1987). Figure 19b illustrates techniques for retaining riffles created by encroachment of vegetation or localized deposition on one river side only. Cross-section 'B' shows an approach which involves lowering both the riffle and the area of deposition by the same amount. In this way the width of low-flow channel is not altered, and the riffle and its plants and animals are retained. Cross-section 'A' shows an alternative approach which involves not bed lowering, but a little reduction in the width of obstruction.

Riffles can also be created by increasing the variation of the longitudinal bed profile in uniform river sections, as shown in Figure 19c. This method will be appropriate only where the bed material is not soft. Alternative methods of creating riffles include dredging only half the bed width. Capacity is rarely impaired by such an approach since the required freeboard is achieved in the dredged half, forming the riffle, and the wet shelf is drowned out during high flows (Fig. 19d).

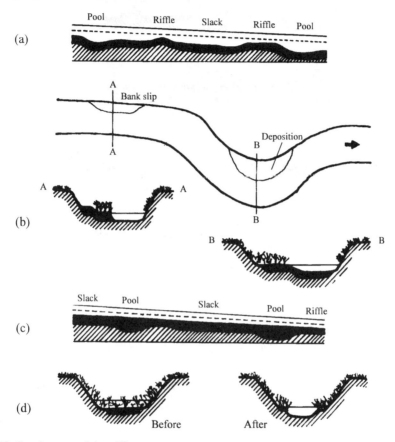

Figure 19. Creating or retaining riffles.

Retaining or creating pools. Pools should also be retained by over-deepening existing ones. Figure 20a shows how existing pools are retained to the same depth, while Figure 20b shows how new pools can be created in a relatively uniform channel.

If bed lowering or removal of movable material is required downstream of a pool, and this would lead to its shallowing and destruction, material from the pool bed should also be removed to an equivalent depth. This ensures that the habitats for plant and animal species associated with pools are retained. It also creates greater stability of the fish population by providing shelter from floods (Brandon 1987).

Creating new pools is more difficult, because the long-term establishment of pools is dependent on fluvial processes which will clean the pool and not deposit silt in it. Selecting the correct location of a new pool is vital. Without such considerations the extra work will result in depressions in the river bed, which act as silt or gravel traps. Where only silt is dredged from a river, there is no reason to create pools since these will be lost very quickly.

Retaining or creating channel profile variation. In rivers with uniform cross-sections, the communities are likely to be very similar, while those with variable bed profiles will exhibit greatest diversity. Figure 20 shows how channel variation can be retained or created: (a) and (b) illustrate longitudinal, and (c) and (d) cross-sectional variations.

Uniform sections tend to suffer from excessive weed growth, requiring a management very soon after dredging. These weeds are usually invasive and exclude other plants, thereby severely affecting fish. On the other hand, a variable channel has a more balanced ecology which requires less management.

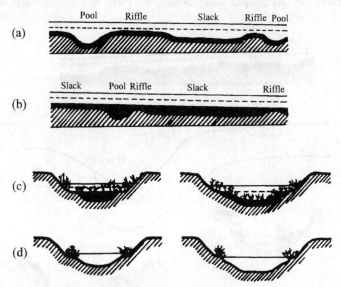

Figure 20. Retaining or creating variation in channel profile.

Leaving the margin undredged. Excepting cases of serious reductions in capacity, maintenance dredging should always retain an underwater strip of vegetation at the bank toe (Brandon 1987). Even the narrowest strip is of ecological benefit. It retains most of the plants and animals, enabling them to re-establish themselves after the works. Figure 21 illustrates retention of marginal edges in narrow and wide rivers.

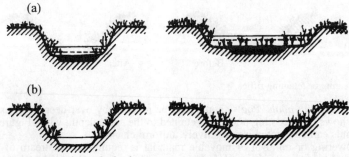

Figure 21. Leaving river margins undredged.

Selective retention of plants and communities. Ecological surveys prior to dredging should identify where rare plants occur within the river section to be managed. These should be ideally left or, alternatively, removed during the dredging and returned after the works. Figure 22 illustrates how, even in a relatively major dredging operation, some vegetation can be retained: (a) individually rare plants left, (b) odd patches left, and (c) narrow strips left undredged. This latter approach is possible only where the substratum is relatively soft, while the former is possible for streams with relatively stable, clay bed.

3.4.4 *Bank cutting*
Since tall plants are used as shelter by birds or mammals, cutting should be done only along

selective reaches so that the animals can find cover. Sensitive management of marginal vegetation involves preserving existing plants where possible. The cutting along the channel thalweg is recommended, leaving bars along the banks, which can be used as a food source and shelter for macroinvertebrate and fish communities.

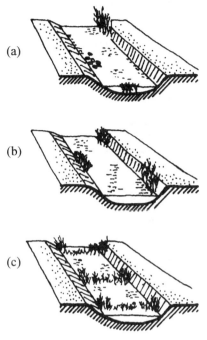

Figure 22. Retaining vegetation during dredging.

Further a few possibilities follows:

1. Reducing frequency of mowing. Due to a combination of financial considerations and a desire to have greater regard for the environment, several authorities have reduced the frequency of bank mowing. In many areas a previous regime of four cuts a year has been reduced to a single cut, with quite positive effects.

2. Leaving strips of bank uncut. Where there is sufficient capacity to allow for a small part of the bank to be left uncut, several approaches have been tried. The simplest involves leaving a strip of uncut vegetation either at the top or at the bank bottom (Fig. 23). Both systems are valuable for creating bankside habitats for invertebrates, birds and mammals which would not otherwise utilize the river if it was mown from top to bottom.

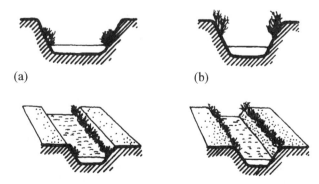

Figure 23. Leaving some bank vegetation uncut.

3. Selective patch retention of vegetation. Figure 24 illustrates an alternative method of leaving parts of a bank unmown. It is selective retention on either an area or species basis.

4. Cutting alternate banks in rotation. This approach means a cutting only one bank in an year. The following year the opposite bank is cut, and the one cut the previous year left untouched. The technique enables the retaining a good root structure on the banks. It also retains plant richness, stops invasive plants to dominate the community, and provides cover for animals.

5. Allowing natural vegetation succession on wide banks. Where very wide banks occur, so that the upper limits of the bank are not critical in determining channel capacity, the banks can be left or encouraged to develop naturally. In the absence of trees and shrubs, a planting should take place.

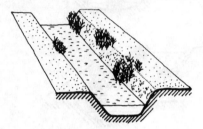

Figure 24. Leaving small areas of bank vegetation uncut.

3.4.5 *Watercourse shading*

An alternative to conventional methods of mechanical cutting and chemical control of aquatic plants is the addition of overhanging marginal vegetation which provides a shading effect. The biomass in naturally shaded channels is significantly less compared to adjacent open sections.

On very small streams the banks should be left ungrazed by livestock, allowing herbaceous plants and grasses to grow. Management is required every 3-5 years to limit plant development. Larger streams require bushes or small trees on the south bank or larger trees on the north bank if the flow is from east to west.

4 CONCLUSIONS

Man-made disturbance of a river ecosystem, caused by traditional bank protection works and river training, creates a plenty of effects, influencing the animal and plant system communities:

1. The vegetation removal appears to be the strongest shock for the system - allochthonous organic particle input is reduced or eliminated, the channel is opened to direct solar input and essentials for autochthonous communities are developed. As a consequence, production relative to respiration is increased, and structural diversity and the system stability are decreased.

The macroinvertebrate fauna is extremely disturbed, according to a loss of macrophytes as a benthic invertebrate habitat. Fishery losses are attributed not only to vegetative cover destruction, but also to pools and riffles elimination, increased turbidity and reduced spawning areas. Widespread destruction of natural plant communities may occur so that the habitats of birds and mammals may be destroyed by the trees, shrubs and grasses removal during construction, thereby affecting these vertebrate species.

2. A sedimentation at downstream reaches, due to the river engineering works, simplifies the substratum and reduces species diversity. Sediment can adversely affect fish eggs, particularly where deposition of silt is excessive. Deposits may blanket stream bed portions, thus eliminating potential spawning grounds and reducing the available food by killing bottom-dwelling organisms.

Downstream siltation due to erosion both during and after construction may destroy pools and riffles, and affect the fish populations. Suspended material may also smother the leaves, thereby preventing photosynthesis, or may be deposited above gravel causing plants to root in unstable conditions, or may lead to the burial of aquatic plant communities.

3. Time of recovery of the original sinuosity can vary from a few years, when the substratum is relatively fine-grained, to thousands of years when the substratum is hard bedrock, depending on the time necessary for a stream to become reinhabited by its original species.

On the other side, the natural river engineering is based on the principle of preservation of natural channel patterns: a meandering water course with pools, riffles, and point bars, as well as tree and shrub growth along streams. The preservation of the dynamics of meandering rivers is hard to fulfil in urban areas. However, in rare cases it is possible to more or less leave rivers to their own devices.

Whatever bank stabilization measures are applied, they should all have a common feature - the retention of the habitats of all aquatic and terrestrial organisms. Transverse structures should be designed in such a way that the aquatic fauna migration in both upstream and downstream direction is not impaired. Regarding the embankments siting and structure, the safety must be the initial consideration, but it is also possible to create new meadow habitats by providing broad untouched areas in front of the structures. This objective is especially difficult to fulfil in built-up areas, but even here it is possible to create small natural stretches.

The natural river engineering procedures and techniques can be divided into a few groups. Floodplain approaches are applied where a whole river corridor can be utilized for an environmentally sound regulation. Other methods discussed (biotechnical engineering and some unique designs from particular areas) comprise a wide range from the local improvements, including selective maintenance, to the large-scale actions. Anyway, all those methods are both ecologically and aesthetically favourable, so that the final solution choice will be made by taking the engineering and economic aspects into consideration.

The stream heterogeneity lies in the base of all the techniques. From the mathematical point of view, in that way, conditions of a numerical calculation are significantly complicated (namely, water level, high flow propagation, etc.), so that a suggestion for further exploration in this domain is the top-quality defining of boundary conditions and the calculation parameters (hydraulic roughness, geometric features of the river bed, etc.). Just in this way, possibilities for a satisfactory numerical calculating are created, and with high-quality results, reliable choice of technically satisfactory variants can be fulfilled.

REFERENCES

Amos, W.H. 1970. *The infinite river. A biologist's vision of the world of water.* New York: Random House.
Barnes, J.R. & Minshall, G.W. 1983. *Stream ecology. Application and testing of General ecological theory.* New York: Plenum press.
Boon, P.J., Calow, P. & Petts, G.E. 1992. *River conservation and management.* Chichester: John Wiley & Sons.
Brandon, T.W. (ed) 1987. *River engineering - part I, Design principles.* London: Institution of Water and Environmental Management.
Brandon, T.W. (ed) 1989. *River engineering - part II, Structures and coastal defence works.* London: Institution of Water and Environmental Management.
Brookes, A. 1988. *Channelized rivers, Perspectives for environmental management.* Chichester: John Wiley & Sons.
Coppin, N.J. & Richards, I.G. 1990. *Use of vegetation in civil engineering.* London: Butterworth.
Gore, J.A. 1985. *The restoration of rivers and streams, theories and experience.* Boston: Butterworth publishers.
Hara, Y. 1993. Erosion Control Facilities Harmonized with their Environment. *Journal of Hydroscience and Hydraulic Engineering.* Japan Society of Civil Engineering. SI-4: 209-217.
Hynes, H.B.N. 1970. *The ecology of running waters.* Liverpool: University Press.
Komatina, D. 1992. *River renaturalization.* Delft: University of Technology.
Komatina, D. 1996a. Environmental impacts of bank protection works and river training. *Proceedings of the International Conference on Environmental Geology and Land Use*

Planning, Grenade, 22-25 April 1996. 1: 231-243.
Komatina, D. 1996b. Procedures and techniques of natural river engineering. *Proceedings of the International Conference on Environmental Geology and Land Use Planning*, Grenade, 22-25 April 1996. 1: 283-305.
Komatina, D. 1996c. Use of natural materials in river training. *Proceedings of the International Conference "Architecture and Urbanism at the Turn of the III Millennium"*, Belgrade, 13-15 November 1996. Belgrade: Faculty of Architecture, University of Belgrade. 2: 365-370.
Komatina, D. 1996d. Environmentally sound river maintenance procedures. *Proceedings of the International Conference "Architecture and Urbanism at the Turn of the III Millennium"*, Belgrade, 13-15 November 1996. Belgrade: Faculty of Architecture, University of Belgrade. 2: 371-376.
Petersen, M.S. 1986. *River engineering*. New Jersey: Prentice-Hall.
Townsend, C.R. 1980. *The ecology of streams and rivers*. Studies in Biology, 122. Southampton: The Camelot Press.
Whitton, B.A. 1975. *River ecology*. Oxford: Blackwell.
Whitton, B.A. 1984. *Ecology of European rivers*. Oxford: Blackwell.

Impacts of natural salt transport in a coastal Mediterranean environment

J. Bach, J. Trilla & R. Linares
Unitat d'Hidrogeologica, Universitat Autònoma de Barcelona, Bellaterra, Spain

ABSTRACT: Under the general framework of a salt balance in a coastal environment, it is the scope of this work to present the methodological issues and corresponding results derived from the quantification of the amount of salts carried in the wind.

By using of several aerosol traps, it is estimated that in a 1 km wide stretch, between 300 m and 1300 m, far from the coastal line, total chloride input is of about 50 kg/ha. Sodium accounts for about 10 kg/ha. These figures were obtained after an experimental period of one year in the Plana of Alt Empordà (Girona), in the northeastern corner of Spain.

1 INTRODUCTION

Salt inputs into soils as well as surface and ground water represents a type of impact on earth surface characteristics, which is often not contemplated in impact assessments.

These inputs can be due to a some human activities, such as certain chemical industries, salt mining or salt pans.

Increased salt supplies can constitute a significant factor in the degradation of soil and water quality.

Salt deposition onto the ground surface can be produced by rain or by direct wind transport. The latter process is probably more important in Mediterranean and other semi-arid environments.

A determination of the amount of salts transported by the wind and the distance of transport from know origin (the sea being an obvious one) can thus help to better assess the impacts of this process, be it from natural or human-created sources.

Usually, the amounts of salts coming in from the sea are assessed by analyzing salt content in the water. The work of Junge & Gustafson (1957) is a good example of this practice. These authors mention that chloride concentration in rainwater ranges in the interval 0.1 - 20 mg/l, showing a slight tendency to higher values in winter months and a decrease in concentration as distance increases from the coastal line. Further inland, those authors observed that chloride concentrations in the rainwater were quite homogeneous and generally smaller than 0.5 mg/l. This concentration was independent of the amount of rain; that is, it reflects the natural salt concentration in the troposphere.

Other authors (Custodio et al. 1985) found chloride concentrations of 9 - 49 mg/l in rainwater samples collected on the roof of the "Escuela Politécnica de Barcelona", 6500m from the coastal line. Nevertheless, these data look anomalous and may reflect atmospheric contamination. Reisman & Ovard (1974) evaluated the influence of wind velocity on salt concentration in the air. They found that the closer to the coastal line and the higher wind velocity, the higher the salt content in the air. The obvious counterpart, decrease in salt concentration going away from the coastal line or with low wind velocity, was also found.

In our case, our objective was to determine the input of salts in areas close to the coastal line. Care was taken to make sure that samples were not contaminated from salts carried by rainwater, to assess salt inputs independently of the amount of rain.

The study was carried out in the Plana de Alt Emporda, Girona, at the northeastern corner of Spain (Fig. 1a). This is a coastal plain with very gentle slopes, usually less than 0.2%, and very low elevations (slightly bellow sea level to 10-15 m several kilometers inland). The plain is formed by Quaternary fluvial deposits, mainly deltaic, somewhat modified by wave action (Bach 1986).

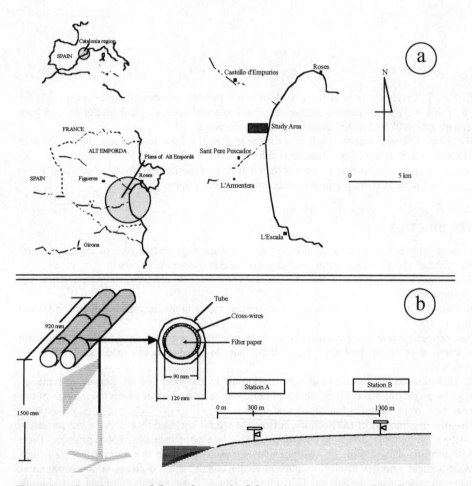

Figure 1. a) Location map of the study area. b) Schematic representation of the "aerosol traps" and location of control stations with respect to the coastline.

2 METHODOLOGY

A modified version of the device used by White & Turner (1970) for the evaluation of the nutrients trapped on the canopy was specially built. This tool is presented in Figure 1b. It is simply a pair of aluminium tubes, 92 cm long and 12 cm in diameter. A 9 cm filter paper

(Whatman No. 50) is placed on nylon cross-wires towards the end of the tube, according to wind direction.

These tubes are placed at each side of a windvane and raised 1.5 m from the ground surface, allowing to whole device to rotate around a fixed base self orienting to the wind direction. Using two tubes two samples for each control point are obtained at every control time.

One of these devices was placed 300 m. (station A) far from the coastline. The other was placed 1300 m inland (station B). Both, A and B were at an altitude between 2 - 3 m above sea level. The filter papers were replaced once every month during two years, from May 1982 to May 1984. The samples were washed out with 100 ml of distilled water for at least 3 hours. 50 ml of the final solution were used for chloride analysis using the Mohr method. The remaining 50 ml was kept for sodium determination using atomic absorption. A blank, made of a paper filter not exposed to the wind and washed with the same amount of distilled water, was always analyzed and results subtracted from samples.

3 RESULTS

The analytical results are expressed as mg collected in each filter paper, abbreviated as mg/disk.

3.1 *Chloride inputs*

Chloride contents in the samples range between 0.7 - 19 mg/disk for the A station and between 0 - 13 mg/disk for the B station, for the whole study period.

A detailed monthly distribution is presented in Figure 2a. A seasonal distribution pattern can be clearly observed in the figure, with highest chloride values in spring and fall and lowest values in winter and summer. Particularly, it can be noticed that the maxima are simultaneous in both stations in the months of November and April. Minima can be located at the months of August and September 1982 and February 1984.

Station A shows, for most months, higher chloride values than station B (Fig. 2b). Moreover, this difference increases when chloride contents are higher in station A. This shows that the chloride source is the sea. Values for station B are higher than those in station A only when the latter are lower than 4 mg/disk.

Data on wind direction and velocity for the experimentation period have also been compiled. These data were taken from the Estartit meteorological station, 20 km south from the testing area, also on the coast. Unfortunately there is no meteorological station in the study area. Nevertheless, given the very similar geographical conditions and its relative proximity, data from Estartit can reasonably be accepted for the Alt Emporda.

Available wind data are in the form of number of days in a month (frequency) in which windspeed equal or above 3 in the Beaufort scale has been registered. These frequencies are grouped according to 8 possible directions. The number of hours with windspeed equal or above 6, with indication of the main direction, is also registered. Some other data, like number of stormy days and their main wind direction are also recorded.

For the testing period, the maxima for number of days with windspeed above or equal 3 are November-December 1982, April-May 1983 and January-March 1984. Some of these months fit with the maxima for chloride concentrations in our sampling stations (Figure 2b). This is the case of November 1982 and April 1983. For some other months this relation is not so clear. Therefore, it seems that the frequency of number of days with velocities equal or above 3 has an important influence on the amount of chloride carried by the wind. Nevertheless, this is also influenced by wind direction.

For the testing period, the maxima for number of days with windspeed above or equal 3 are November-December 1982, April-May 1983 and January-March 1984. Some of these months fit with the maxima for chloride concentrations in our sampling stations (Fig. 2b). This is the case of November 1982 and April 1983. For some other months this relation is not so clear. Therefore, it seems that the frequency of number of days with velocities equal or above 3 has an important influence on the amount of chloride carried by the wind. Nevertheless, this is also influenced by wind direction.

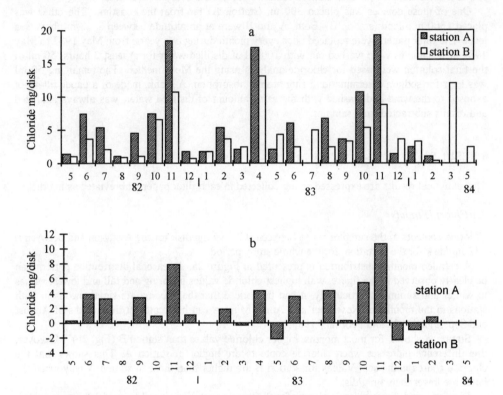

Figure 2. a) Contents of chloride in the filters of the "aerosol traps", in mg/disk, for both stations.
b) Difference in the values for chloride between station A station B. Top values for periods during which A>B; bottom values viceversa.

Figure 3 presents histograms of the number of days per month with the 8 wind directions considered (N, NE, E, SE, S, SW, W, and NW). It can be observed that the predominant wind directions are those with North and South as main components; in local terminology "tramuntana" and "mitjorn", respectively. Both are frequent mainly in the winter and spring seasons; that is, seasons which do not show maxima in chloride carried by the wind.

A closer look at Figure 3 shows that the eastern wind, locally called "llevant", shows a good correlation with maxima in chloride concentration; months of November and April. Other directions do not show any correlation with chloride concentration. A closer look at Figure 3 shows that the eastern wind, locally called "llevant", shows a good correlation with maxima in chloride concentration; months of November and April. Other directions do not show any correlation with chloride concentration.

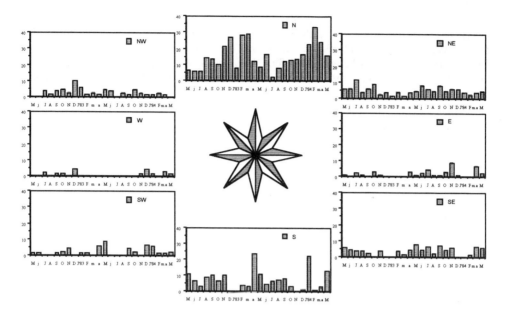

Figure 3. Frequency (%) of winds with velocity equal or above 3 in the Beaufort scale, according to the eight general directions, during the experimental period.

Moreover, if we look at the distribution of the number of hours with windspeed equal or above 6, for winds with a main eastern component and stormy days (Fig. 4), we can also notice some correlation with chloride distribution.

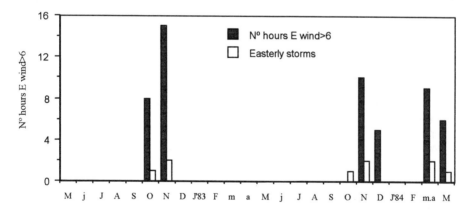

Figure 4. Distribution of number of hours with windspeed equal or above 6, for winds with a main eastern component, and stormy days, during the experimental period.

Figure 5 presents the results of the correlations established between chloride content in the samples and frequency of wind velocity above 3 or easterly storms, for both sampling stations.

The calculated correlation coefficient between chloride concentration and number of days of easterly winds is 0.69 and 0.65 for A and B stations, respectively. The correlation with the number of days with easterly storms is 0.77 and 0.6 respectively. That is, the correlation is always better for station A, closer to the sea, than for station B, for both stormy days and number of days with easterly winds of velocity 3 or higher. These data could be refined if wind velocity, not only frequency, could be accounted for. Nevertheless, it seems clear that salt inputs are mainly due to winds with an eastern component. Given the geographical situation of the test area this was to be expected.

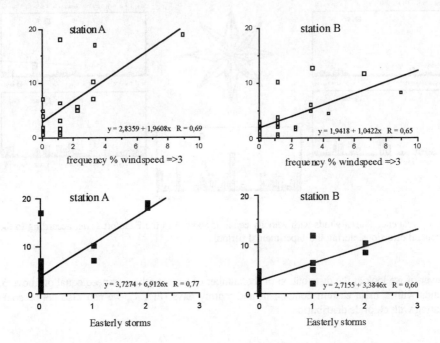

Figure 5. Correlation between chloride contents in the aerosol traps and easterly winds (top), and easterly storms (bottom), for stations A and B.

In the light of the data discussed above we can confirm our initial assumption that the main source of chloride collected in the samples is the sea and that is carried mainly by easterly winds. These results are similar to the ones obtained by Jansa et al (1988) in Menorca, for "tramuntana" northerly winds. The fact that station A shows higher contents than station B is coherent with that interpretation.

It can be inferred from the data show in Figure 2 that there is a considerable difference in salt deposition according to distance from the sea. This difference is roughly 25-50% for most months. That is, salt deposition on the soil by wind transport is significantly decreased for distances greater than 1 km from the source.

The fact that some months, as described above, show higher chloride deposition in station B can probably be explained by salt mobilization from the soil surface.

Estimates of the impact of chloride deposition on the soil can be made from the data above. An extrapolation of the amounts deposited on the filter paper gives values between 0.5 kg/ha and 28.5 kg/ha per month. The total deposition in a year, for the period studied, would be about 109.6 kg/ha and 68 kg/ha for stations A and B respectively.

These values should be interpreted as minimum values, because it is unlikely that the sampling device has a 100% efficiency trapping the salt carried in the air. That is, the process of salt transport by the wind can be quite significant, especially with certain wind conditions.

3.2 Sodium inputs

The methodology followed for sodium sampling has been the same as the one for chloride. Experimental data obtained in both stations are plotted in Figure 6a. For station A sodium concentration ranges in the interval 0.13 - 4.18 mg/disk per month, while for station B the values are 0.01 - 3.58 mg/disk per month.

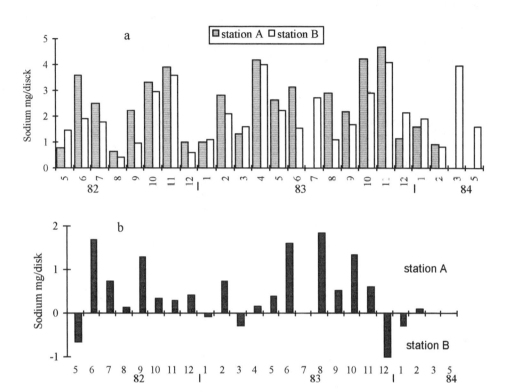

Figure 6. a) Sodium contents (mg/disk) in the "aerosol traps", for both stations. b) Difference in the values for sodium between stations A and B. Top: months with sodium higher in A than in B; bottom: viceversa.

Sodium maxima follow the same distribution as chloride maxima. Nevertheless, differences between extreme values are not as noticeable as for the latter. The same can be applied when comparing data from both stations. In general, differences are not greater than 2 mg/disk for the whole set of samples (Fig. 6b).

A more detailed look at the data shows that for some 5 months, sodium content registered in station B is higher than content in station A (Fig. 6b). In three of these months the same type of relation is observed with the chloride data. The other two months show an opposite relationship, but they are months of low sodium deposition.

As sodium data follow the same trends as chloride ones, the interpretation presented for the latter can also be accepted for the former.

Figure 7 shows the correlation between sodium contents and easterly winds variables for each station. The resulting correlation coefficients are similar to those obtained for chloride. They are 0.64 and 0.74 for the correlation sodium content/number of days of easterly winds, for stations A and B, respectively, and 0.62 and 0.7 for sodium content/number of days with easterly storms. Notice that correlation coefficients are higher for station B than for station A. Nevertheless, differences are not high; the range between extreme correlation coefficients (0.62 - 0.74) is almost identical to the one obtained in the case of chloride. These coincidences indicate that the sodium source is also the sea, and that the easterlies are the main input vector.

These monthly values are in the same range of the results obtained by White & Turner (1970); using paper filters 5.5 cm in diameter, these authors registered sodium contents in the range 0.1 - 3 mg/disk, exceptionally 5 mg/disk and 13.63 mg/disk, when wind velocity was above 15 km/h.

Following the same procedure used for extrapolating chloride area deposition on the soil from disk data, we can estimate that sodium accumulates inland at rates in the range 0.016 - 6.6 kg/ha per month depending on the conditions. On a yearly basis, for the same period May 1982 - April 1983, values are 33.1 kg/ha for station A and 25.5 kg/ha for station B. Because of the same reasons mentioned in the case of chloride accumulation on the soil, these values have to be considered as minima. Real values should also be higher.

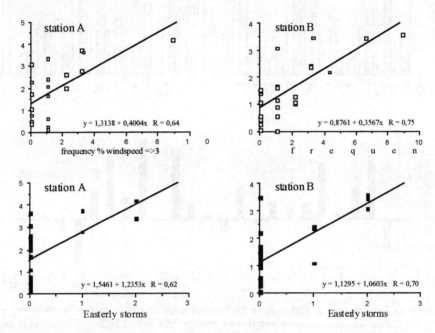

Figure 7. Correlation between sodium contents in the aerosol traps and easterly winds (top), and easterly storms (bottom), for stations A and B.

3.3 Ratio Cl^-/Na^+

If, as explained above, the source of salt deposited in our devices is the sea, it should be expected that Cl^-/Na^+ ratios in the sampling devices be similar to that of seawater; that is 1.8. However, this is not exactly the case.

Figure 8 shows the Cl^-/Na^+ ratio for those months during which Cl^- deposition was greater than 2 mg/disk (to avoid noise effects in the data). It is clear that the ratio is, in general, close to that of seawater, but always somewhat higher and in some months significantly higher. The ratio is also quite similar for both stations, although the values are normally higher for A than for B.

The months during which the Cl^-/Na ratio is above 3 are those with higher Cl^- deposition; that is, the ones with intense easterly winds. This shows that wind transport is more efficient for Cl^- than for Na^+.

These findings are somewhat contradictory with the results of former work. Rossby & Egnér (1955) also found Cl^-/Na^+ ratios around 3 for rain water during southerly storms in northern Europe, whereas Jungle and Werby (1958) found that rainwater had ratios below the normal seawater value.

Nevertheless, it must be kept in mind that our data do not correspond to rainwater, but to wind transport and deposition.

We have no satisfactory explanation for this apparent anomaly. Perhaps some ion fractionating or addition of chloride from the soil surface are taken place. More work is under way to elucidate this point.

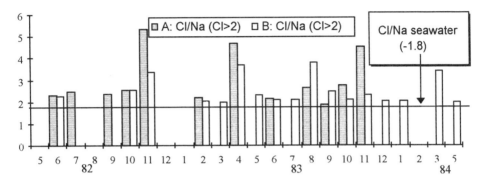

Figure 8. Cl^-/Na^+ ratios in samples collected during months with $Cl^- > 2$.

CONCLUSIONS

The results presented show that wind is a significant factor of salt transport and that the sea is the main source in the area studied. Nevertheless, it appears that direct mobilization from the soil surface also plays a role.

Chloride and Sodium impacts into the soil as a result of this process can be estimated to be around 100 kg/ha/year and 30 kg/ha/year, respectively. This represents an important "natural impact" on soil and surface and groundwater.

The relatively high Cl^-/Na^+ ratios found in the samples suggest that ion fractionation or selective chloride mobilization from the ground surface might be playing a role too.

The differences observed between the two sampling stations indicate that, as should be expected, distance is an important factor in the reduction of these effects. The amount of salt can decrease about 25-50% in 1 km. The data also indicate that in the case of activities which imply exposure and/or mobilization of salts at the ground surface, significant impact ought to be expected at distance of at least 1-2 km downwind.

REFERENCES

Bach, J. 1986. Sedimentación holocena en el litoral emergido de "L'Alt Empordà" (NE de Catalunya). *Acta Geológica Hispánica*, 21-22: 195-203.

Custodio, E.; Pelaez, M.D. & Balagué, S. 1985. Datos preliminares sobre la aportación mineral por la precipitación atmosférica. *Tecnologia del Agua* 18: 51-56.

Jansa, A.; Rita, J & Calafat, A. 1988. La salinización de origen eólico en Menorca (Baleares). Primeros datos experimentales e interpretación. *Comunicaciones de la V^a Asamblea Nacional de Geodesia y Geofísica. Ed. Inst. Geográfico Nacional.*

Junge, T. & Gustafson, J. 1957. On the distribution of sea salt over the United States and its removal by precipitation. *Tellus* IX 2: 164-173.

Jungle, C. E. & Werby, R. T. 1958. The concentration of choride, sodium, potassium, calcium, and sulfate in rain water over the United Satates. *Journal of Meteorology*. 15 (5): 417-425.

Reisman, J. I. & Ovard, J. C. 1974. Efecto de las torres de refrigeración sobre el medio ambiente. *Ingeniería Química* 61: 131-138.

Rossby, C. G. & Egnér, H. 1955. On the chemical climate and its variation with thw atmospheric circulation pattern. *Tellus* 7: 118-133.

White, P. & Turner, J.J. 1970. A method of estimating income of nutrients in a cath of airbone particles by a woodland canopy. *J. Appl. Ecol.*, 7: 441-461.

Environmental reclamation of dismissed quarries in protected areas: A case study near Rome (Italy)

G. Di Filippo, M. Pecci & F. Silvestri
ISPESL - DIPIA (Productive Settlements and Environmental Interactions Department) Roma, Italy

F. Biondi
Geologist, Roma, Italy

ABSTRACT: Abandoned quarries and open pit mines represent a problem in environmental assessment and for landscape too, above all in natural park and in protected area. The approach to the study must be managed in a multidisciplinary way, not only by the law-making, but also from the technical-scientific point of view. In Italy legislative system goes on slowly, also because of the difficulty in laws application themselves. In fact State, Regional and Municipal Authority issue laws often similar and sometimes non homogeneous. So studies and researches on these topics need a suitable technical-scientific support. In this paper the authors tried to approach the study starting from the scientific literature and from tele-detected data, going on through controls on the field (morphological, geological-technical and geoelectrical surveys) and ending with the proposal of hypothesis of solution. The study has been carried out in an abandoned quarry North of Rome, located in the Archeological Park of Veio. The aim is to suggest solutions to the environmental reclamation of the area, according to previous technical reports too.

1 INTRODUCTION

Quarry activities, distributed wide spread all over the countries, generally cause problems of environmental compatibility.

The effects of a quarry/open pit mine may be distinguished in transitory and permanent: transitory effects, such as acoustic pollution, vibrations, ash atmospheric pollution, heavy traffic (Silvestri & Tagliaferro 1993) strictly depend on working activities, generally ending with the end of activities themselves; permanent effects (equilibrium alteration in some environmental components such as hydrology, hydrogeology, slope stability, landscape and vegetation) may remain after the quarry decommissioning, if no rehabilitation has been carried out (Silvestri 1988).

So it is possible to notice the greatest and, often, the worst effects on the landscape, particularly in the case of quarry along the slope, where debris and wastes are stored up with no order and with no reclamation plan, often causing slope stability problems. About these topics, it seems noticeable the exploitation on one steep wall perpendicular to natural slope, the wrong location on stratified geological formation, the erosion at the toe of the slope, the non suitable usage of explosive.

According to the lithological characteristics and to the mechanical behaviour, the following risk conditions are active in the area:
  1. toppling and fall on rock masses;
  2. sliding and slumping on earths.

Landslide risk is overburden by variation of water table, inducing the mechanical characteristic decay in earths (Silvestri 1988).

In Italy the problems linked to the reclamation of dismissed or abandoned quarry areas represent a steady increase emergency. In fact nowadays Italian laws order everywhere, in planning phase too, the site reclamation, but it has been not so, until few years ago. So there are many quarries, abandoned in the last years, without environmental restoration plan. In some areas the problem has a great evidence because of the presence of precious rocks. A case study has been recently presented in the exploitation area of the so called "Pietra di Trani", near Bari (Southern Italy). In this occasion ISPESL-DIPIA (Higher Institute for Occupational Safety and Health - Department of Productive Plants and Interaction with the Environment) has been required to carry out a study to assess the characterization, the evaluation and at the reclamation of territory. In the case of Minervino Murge, a little town in the surrounding area of Trani (Puglia Region), 12% of the municipal territory has been interested by quarry activities (Di Filippo & Silvestri 1994).

2 QUARRY ACTIVITIES PLANNING: STATE OF THE ART IN STATE AND REGIONAL LAWS

The main legislative source in Italy is represented by R.D. (King Act) 29/07/1927, n. 1443, so called "mining law", successively modified with law 06/11/1941, n. 1360. At the beginning of the seventies State jurisdiction has been passed to the regional Authority in two phases, corresponding to DPR (President of the Republic Act) 14/01/1972, n.3 and DPR 24/07/1977, n. 616. In the first phase the regional Authority acquired overseeing functions and the management of mining concession, in the second phase care and inspection powers, coupled with the safety controls.

In this way quarry and open pit mine activities have been included into the planning and development programs, with the arrangement of P.R.A.E.(Regional Plan of Extractive Activities).

In Lazio region extractive activity is ruled by regional law 05/05/1983, n. 27 "Norme per la coltivazione delle cave e torbiere della regione Lazio " ("Rules about quarry and pit-bog working in the Lazio Region"). The law, in the case of the research and the exploitation of quarry, obliges to present a study of EIA (Environmental Impact Assessment) and a Safety Plan, the last one required for active quarries too.

3 MORPHOLOGIC, GEOLOGICAL AND HYDROGEOLOGICAL OUTLINE

Studied area is represented by a quarry (Fig. 1), located in the site named "Monte Aguzzo" near Formello (about 2 kilometres), a little town in the northern country side of Rome.

Figure 1. View of the quarry area with wastes at the bottom of the quarry and sliding on the right side.

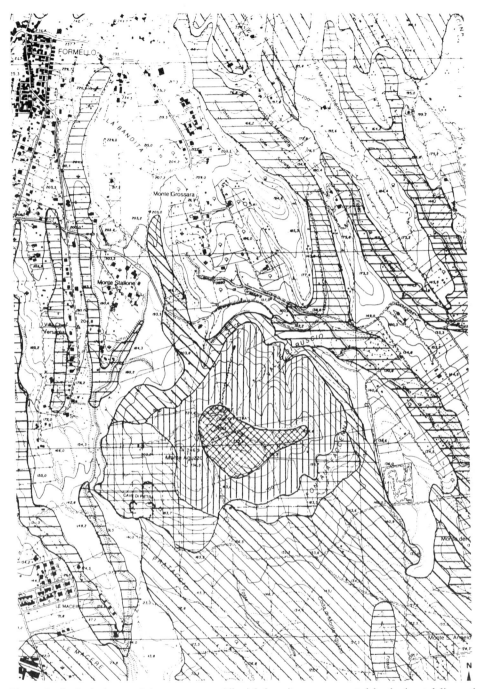

Figure 2. Geological map of the quarry area. Alluvial deposits are represented by horizontal lines; the mainly leucitic scoriae cone (quarry area) is represented by a oblique grid; the mainly leucitic compact lavas are represented by a upright grid; white cineritic tuffs (explosive phase of Eastern Sabatino Volcano) are represented by vertical lines; grey tuffs (explosive phase of Western Sabatino Volcano) are represented without any symbol; trachytic ignimbrites, often lithoid and sometimes stratified are represented by oblique lines.

From the morphologic point of view the main element is represented by volcanic cone of Sacrofano, reaching 370 m asl. Hydrographic network in fact has a typical "radial pattern", with deep gorges and steep wall-sided valleys in volcanic-tufaceous deposits.

Particularly southern valleys start from the volcanic structure with narrow shape and open themselves toward Campagna Romana, with characteristic plate bottom and meanderings.

Monte Aguzzo (247 m asl) is a little hill and is located about in the middle of the slope of the volcanic structure, gently dipping from the highest elevation to Prima Porta (30 m asl), a suburb located few kilometres north of Rome along the ancient "Flaminia" road. It is characterized by a typical horse-shoe shape, northern oriented; the top is cutted by a split, southward opening and transforming itself in a little stream (Fosso di Monte Aguzzo).

From the stability point of view it is possible to observe only localized superficial soil slope deformation, generally due to natural actions. The area belongs to the volcanic Sabatino District and Monte Aguzzo represents a secondary eruptive centre, characterized by a rapid evolution, with moderate magma overflowing and by the construction of a scoriae cone. Piroclastites and particularly ignimbrites are the most frequent units outcropping in the area; instead lavas outcrop only in a restricted area along southern slope of Monte Aguzzo. Alluvional deposits outcrop in restricted areas along more wide valleys.

In Figure 2 the geological setting of the studied is shown: evidences of faults are not detected in surrounding areas, but into the quarry it is possible to survey a great number of joints along artificial slope (see geological-technical survey), with a preferential (excluding data related to the three main artificial walls) Meridian, East-West, Apennine and Antiapennine directions, strictly linked to cooling processes and tectonic actions too. Particularly it is possible to hypothesize that vertical and sub-vertical systems are related to the first process phenomenon and more gently dipping systems to the second one.

It is also important to remember that Formello and surrounding areas have no seismic classification (MINISTERO LAVORI PUBBLICI - CONSIGLIO SUPERIORE DEI LAVORI PUBBLICI - SERVIZIO SISMICO, 1986).

From the hydrogeological point of view the hydrographic network includes Fosso della Pietrara (West and North-West of Monte Aguzzo), Fosso della Grossara (North of Monte Aguzzo), Fosso dell'Acqua Viva (East of Monte Aguzzo) and Fosso di Monte Aguzzo (in the southern area): they are all torrent-like rivers, above all in the upper part, joining together near Prima Porta in Fosso della Valchetta, right tributary stream of Tiber River.

In Figure 3 an hydrogeological scheme (about 1:27,500 scale) of the area is shown, where black thin lines represent hydrological pattern, black dashed lines represent isopiezometric lines, big arrows represent hypothesized flow direction of ground water, empty squares represent water fountains, the grid box represents the studied quarry site, the drops represent important spring and finally the circles represent wells. In the area the water springs and fountains (Tab. 1) have been detected by Ventriglia 1989.

A further analysis was carried out in order to evaluate permeability behaviour: two in situ infiltration test, using field Muntz permeameter, was performed in A and B test point of Figure 4, where obtained values were shown in a table and related diagram. The values indicate high permeability behaviour (K about $1,5 \times 10^{-4}$ m/s) of the outcropping soils, especially at the bottom of the quarry.

Starting from these data, from technical and scientific literature and remembering that the lower quarry level is about 200 m asl, it is possible to hypothesize the presence of the upper water table, with a piezometric level about 25 m from the field and flowing southward directed. Ignimbrites and tuffs have been classified in Ventriglia 1989 with permeability from average to low, while scoriae with high permeability.

Table 1. Springs/fountains morphological features related to the position of the quarry.

Spring/Fountain	Height (m asl)	Distance (km)	Direction	Flow (l/s)
Fontana Faccenna	175	about 0.4	North-East	0.1
Valle di Ruscio	140	about 0.6	North-East	
Fonte di Acqua Palombina	150	about 0.7	East	0.2
Fonte di Acqua Viva	120	about 1.45	East-Southeast	0.5
La Villa (drainage shaft)	180	about 0.7	North-West	0.1
Fontana Rutula	160	about 1	North-West	0.1
Sorgente Passo dello Scannato	105	about 1.5	South-Southwest	0.05
Fontanile Caldarella	75	about 3	South-East	0.1

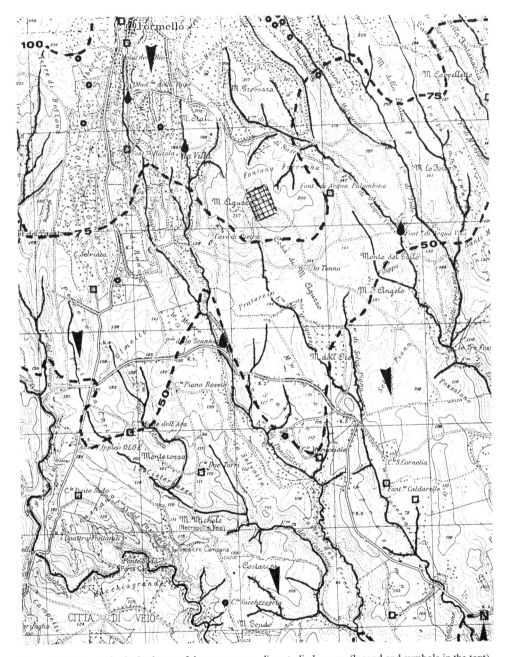

Figure 3. Hydrogeological scheme of the area surrounding studied quarry (legend and symbols in the text).

4 GEOLOGICAL-TECHNICAL SURVEY

In the area of the studied quarry a quick field geological-technical survey according to Bieniawski (1979) and to Amanti & Pecci (1995) has been carried out. Structural data (joints and discontinuities) are shown in the diagram of Figure 5a, coupled with contours of Figure 5b; in Figure 1 a view of the quarry and the stability situation of the artificial slopes and walls is shown. Geological-technical data elaboration (Bieniawski 1979, Amanti & Pecci 1995) has been aimed at characterizing the main quarry slopes, from the mechanical point of view, as shown in Table 2.

	permeability (10^{-4} m/s)			
TEST	1st hour	2nd hour	3rd hour	4th hour
A	2,14	2,01	1,92	1,90
B	2,06	1,65	1,53	1,50

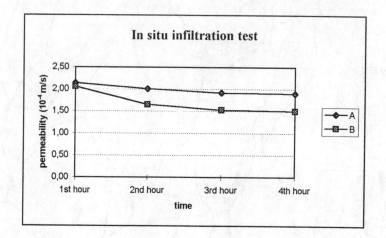

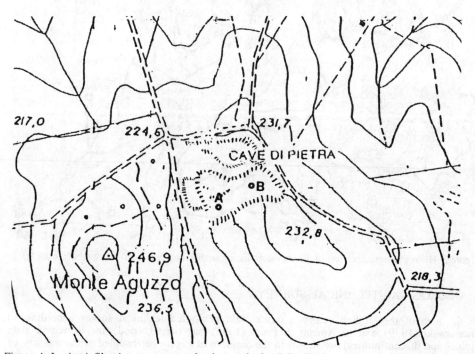

Figure 4. In situ infiltration test: surveyed values and related distribution diagram in the upper part and location of the test points.

For each quarry wall, a fair rock mass quality was found, that should correspond to the cohesion and friction classes recognized in Table 2 (Bieniawski 1979). Volcanic scoriae outcropping in the quarry show ambiguous mechanical behavior; in fact they are classifiable like granular earth, but, at the same time, are interested by joints and discontinuities (Figs. 1 and 5). The behavior induced the authors to hypothesize, the presence of interlocking forces and not only real cohesion forces (hypothetic cohesion in Table 2).

Table 2. Rock mass classification of the geological-technical units in the quarry.

Quarry wall	Wall azimuthal attitude (°)	RMR modified	Rock mass quality	Hypothetic cohesion class (kN/m^2)	Friction class (°)
Eastern	N 50/87	64	fair	200 - 300	25 - 35
Northern	N130/85	55	fair	200 - 300	25 - 35
Western	N200/87	51	fair	200 - 300	25 - 35

So local phenomena of instability, like the sliding/fall of Figure 1, are strictly linked to lithological and genetic characteristics (cooling processes), to the structural setting of the area (Meridian, East-West, Apennine and Antiapennine Systems) and to the mechanical decompression on steep/vertical slope, after the abandonment of the quarry, in residual friction condition (evaluated in about 30 as natural dipping of debris at the toe of the artificial slope).

The phenomenon represents an important element for the evaluation of safety condition and for the environmental restoration of the area.

5 GEOELECTRICAL SURVEY

Two Vertical Electrical Soundings (VES) have been carried out in the area: the first one (VES1) inside the quarry, along azimuthal direction N 50 for a total length of 100 m; the second one (VES2) outside the quarry, along the external border of the same quarry (azimuthal direction N 340) for a total length of 400 m. "Schlumberger quadripole" has been used as investigation method.

Resistivity values detected in both VES have been extremely discordant and non homogeneous; their distribution on bi-logarithmic graphics (Resistivity-AB/2) allows to recognize only restricted intervals of a continuous curve. So it is possible to hypothesize that such a distribution of resistivity values could be linked to lithological variation and not to a regional and continuous water table.

The quarry is located inside a volcanic cone, builded by volcanic scoriae, included stones, compact lavas and cineritic tuffs: resulting setting is extremely variable in space and could justify wide variation in resistivity values. In fact high resistivity values are probably due to current passage into lava thickness, non homogeneous, but included in pyroclastic products, characterized by low resistivity.

6 MORPHOMETRICAL ANALYSIS

Photogeological analysis and aereophoto-restitution have been carried out using areophotographs 1:5000 scale.

Table 3. Morphological features of the studied quarry.

Side	Length (m)	Vertex	Height (m asl)
Northern	about 130	North-West	about 224
Southern	about 175	South-East	about 230
Eastern	about 133	North-East	about 230
Western	about 165	South-West	about 210

Aereophotographs have been elaborated by the analytic photogrammetric device Stereobit and restituted in digital format by GRES Program (Galileo Siscam, unpubl.).

Methodology of digital restitution has been allowed to obtain, at the end of all procedures, points and contours useful for the following elaboration. In such a way morphology of irregular four sided shape (Tab. 3) and features of the quarry (Tab. 4) have been calculated by a mathematic model.

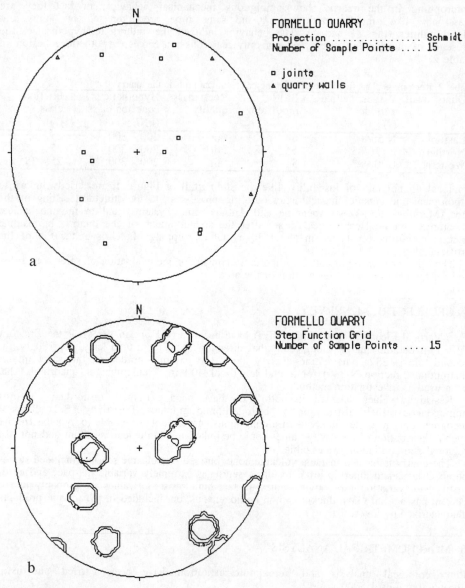

Figure 5. Results of structural elaboration: a) Schmidt and b) percent frequency contour diagrams, both on lower emisphere.

Quarry datum-plane has been referred to real different levels to obtain filling volume of Table 4.

Starting from data of volume, in Table 4 it is possible to estimate the available total volume from the relation:

$$V_{tot} = \frac{1}{3}V_1 + V_2 \qquad (1)$$

The resulting calculated total volume (V_{tot}) is 303,099.91 m^3.

Table 4. Quarry feature and corresponding calculated values.

Quarry feature	Value
total area	26577.98 m^2
main datum plane level	233.43 m asl
main filling volume (V_1)	394889.95 m^3
maximum wall height	33.43 m
minimum level	200.00 m asl
maximum level	233.43 m asl
middle datum plane level	224.00 m asl
middle filling volume (V_2)	171469.39 m^3
secondary middle datum plane level	210 m asl
secondary middle filling volume (V_3)	12543.14 m^3

7 CHOICE CRITERIA AND ELEMENTS FOR THE ENVIRONMENTAL ASSESSMENT

The study, carried out to determine, from the technical and economic point of view, the best rehabilitation of the area, started on the initiative of the Municipal Authority. The aim of the procedures was to provide a general improvement of the environment and a reduction of the present risks. The choice of the different types of rehabilitation, with the re-use of the area, has been analysed following a classification procedure based on a qualitative evaluation of the most significant factors.

The solution achieved has been validated by a comparative analysis based on the effects, comparing the present situation and the new plan. In order to choose the type of re-use it has been stated that the recovery of the quarry area has to consider not only the morphology of the surrounding territory and its previous conditions, but also the socio-economic demands.

The planning solutions have been based on the following kinds of rehabilitation:
1. naturalistic
2. forest-agricultural production
3. leisure
4. technical-functional.

The study methodology assess the reduction of the impacts on the environment, in transforming the area. The chosen procedure has also considered the importance of a full participation of the population in making decisions. A qualitative classification has synthesised the situation. In fact the numeric methods, which may give an outward accuracy, have been avoided, especially in the preliminary phase. In Table 5 a simplified method of qualitative classification of the alternatives is shown, with a range of "high - medium - low", applying to the advantages and in Table 6 the present and the future situation are compared, using a "high - medium - low" range, applying to the impacts.

Table 5. Classification of the re-use/rehabilitation alternatives. Simplified method of qualitative classification of the alternatives.

Selected factors	Rehabilitation typologies			
	naturalistic	forest-agriculture	leisure	techno-functional
Health and safety	medium	medium	medium	high
Land use	medium	medium	medium	medium
Socio-economic parameters	low **	low ***	low **	high
Landscape	high	medium	high	medium
Environmental features*	high	medium	medium	medium

* air, water, soil, rumour etc.; taking in account: ** distance, users, other areas and *** agricultural quality of soil, actual needs etc.

This approach has been an effective tool in order to synthesize the different information and to give to the population real information about the unavoidable compromise solution the decision may involve. All kinds of rehabilitation, except the technical-functional one, have usually very high costs, that do not justify the expected advantages. The technical-functional recovery, on the contrary, is the only one able to assure a general reduction of the risks and impacts, in addition to the economical advantages through a right realization and management of the project. It is important to note that, following the cost-benefit evaluation, the use like landfill for inert wastes (non-toxic) has been considered the best solution in order to obtain the rehabilitation of the area.

Table 6. Comparative analysis between the present situation and the proposed re-use plan.

Effects on:	Present situation (Decommissioned quarry)	Re-use and rehabilitation Plan
Healt and safety	high	low
Land use	medium	low
Socio-economic parameters	medium	None (positive effects)
Landscape	high	medium
Environmental features*	high	medium

* air, water, soil, rumour etc.

To this aim the project, including the Environmental Impact Assessment, has been entrusted to a specialized firm, after an official agreement under the Municipal authority and ISPESL's supervising.

REFERENCES

Amanti, M. & Pecci, M. 1995. Proposta di una scheda per la raccolta e l'informatizzazione dei dati utili alla classificazione e caratterizzazione degli ammassi rocciosi. IV Conv. Giovani Ricer. in Geol. Appl., Ottobre 1995, Riccione. *Quad. di Geol. Appl.*,1/2: 1-8.
Bieniawski, Z.T. 1979. Geomechanics classification of rock masses and its application in tunnelling. Proceed. IV Cong. on Rock Mech, Montreux, A: 27-32.
Di Filippo, G. & Sivestri, F. 1994. Studio degli impianti estrattivi nella provincia di Bari. II Congresso Italo-Brasiliano di Ingegneria Mineraria, 26-27 settembre 1994. Verona.
Galileo Siscam. *Stereobit, manuale d'uso e manutenzione.* Unpubl. Technical Report, Firenze.
MINISTERO LAVORI PUBBLICI - CONSIGLIO SUPERIORE DEI LAVORI PUBBLICI - SERVIZIO SISMICO 1986. *Atlante della classificazione sismica nazionale.* Ist. Poligrafico e Zecca dello Stato, Roma.
Silvestri, F. 1988. L'impatto ambientale degli impianti di cava. *Ambiente e sicurezza*: 16-20, Roma.
Silvestri, F.& Tagliaferro, I. 1993. Problematiche di impatto ambientale nell'esercizio di cave di materiali inerti: inquinamento acustico, da polveri e da vibrazioni. *Fogli di informazione ISPESL*, 1993 (2): 36-44.
Ventriglia, U. 1989. *Idrogeologia della Provincia di Roma - Regione Vulcanica Sabatina.* Amm. Provinciale di Roma, Roma.

Landscape analysis based on enviromental units and visual areas. The use of geomorphological units as a basic framework. La Vall de Gallinera, Alicante (Spain)

I. de Villota
Dpt. de Ideación Gráfica Arquitectónica. E.T.S. Arquitectura. U.P.M. Madrid, Spain

J.L. Goy
Dpt. de Geología. Facultad de Ciencias. Univ. de Salamanca, Spain

C. Zazo
Dpt. de Geología. Museo Nacional de Ciencias Naturales. C.S.I.C. Madrid, Spain

I. Barrera
Dpt. de Biología Vegetal I. Facultad de Ciencias Biológicas. Univ Complutense. Madrid, Spain

J. Pedraza
Dpt. de Geodinámica. Facultad de Ciencias Geológicas. . Univ. Complutense. Madrid, Spain

ABSTRACT: An approach to landscape description and assessment is presented. The method proposed can be used for both environmental impact assessment and land-use planning. The approach includes the consideration of two types of units, homogeneous environmental units and visual basins. Geomorphological domains are the basis for sectoring and organizing the territory for the definition, mapping and description of environmental and landscape units. Two kinds of characteristics of landscape are considered for the assessment: descriptive (quality, singularity) and potential ones (fragility/impact absorption capacity and use aptness). The assessment relies on both objective descriptions and subjective evaluations.

The approach proposed is applied to the area of 'La Vall de Gallinera', SE Spain. The results of the analysis are presented and discussed.

1 INTRODUCTION

Landscape, formerly understood mainly as an aesthetic background, has come to be considered in recent years as yet another type of natural resource. This as such, makes it comparable to other resources: vegetation, soil, fauna, etc., thus requiring protection, conservation and recuperation. Landscape plays an important role in land use, land management and environmental conservation. Due to the consideration of visual landscape as a natural resource, it becomes necessary to establish procedures for describing, assessing and comparing it with other resources, in order to determine its land use capacity and to assess the impacts it may experience. These procedures could be based on relatively 'objective' criteria or explicitly on use of 'subjective professional' judgements. In this analysis, landscape is understood in a wide sense, as a resource of an aesthetic nature; the perception of this resource is determined by the understanding of the environment as a network of interactions between abiotic, biotic and human elements and processes.

One objective of this work is to identify the main factors which determine the present state of the landscape; that is to say which elements, interrelationships and processes are responsible for its present structure and condition. Another objective is to define, map and describe landscape units, which can be used for the assessment of potential impacts and for land-use planning.

As landscape perception is determined to a considerable extent by shapes, volumes, spaces, structures/textures and colours, all of them (mainly the geometry of space) depending to a large degree on geomorphological features. These are also the least variable landscape characteristics. Therefore, it seems logical to use geomorphological domains as the framework for identifying and describing landscape units.

1.1 *Landscape qualities*

The analysis of landscape can be carried out in two ways: a descriptive one, which considers intrinsic characteristics of the landscape, and by its potential use, taking into account functional capacities (Table 1).

Table 1. Analysis of landscape qualities (Villota et al., 1996, modified)

DESCRIPTIVE ANALYSIS	VISUAL QUALITY	GRADIENT OF:	LUMINOSITY, COLOUR, TEXTURE, SIZE
		OUTLINE	
		ASPECT	
		LINE	
		COMPOSITION	
		LANDMARKS	
	SINGULARITY	STATIC	SINGULAR, RARE, UNIQUE
		DYNAMIC	CHANGES OF EXPECTATION
POTENTIAL ANALYSIS	FRAGILITY AND/OR IMPACT ABSORPTION CAPACITY	DENSITY, TEXTURE, LUMINOSITY, VISION ANGLE, COMPOSITION	
	LAND USE APTNESS	DENSITY, TEXTURE, LUMINOSITY, VISION ANGLE, COMPOSITION	
	INTERVISIBILTY AND ACCESIBILITY	POSSIBILITY TO BE SEEN FROM ACCESIBLE POINTS	

Visual quality and singularity are intrinsic landscape characteristics at any given moment in time and can be used for a 'descriptive diagnosis', while fragility to impacts and suitability for a given land use, are considered as potential capacities.

Visual quality is a complex property determined by a series of perceptual variables. Following Gibson's idea of visual invariants (Montes 1992) the basic parameters for the visual comprehension of a reality are the gradients or gradations of luminosity, colour, texture and size, together with the more complex concepts of contour, shape and composition. From a descriptive point of view, these variables are partially intrinsic since they are the expression of the characteristics of every element in the area to be studied, but they are also external and depend on the relative position between the observer and the perceived object. From the point of view of its value, they depend on the observers, their education, culture and sensitivity.

The concept of singularity of a given area has an absolute, static value as well as a dynamic value. A landscape may be described as singular, peculiar, unique or unusual. But singularity is also a relative and dynamic concept that depends on the surrounding conditions, since it refers to something unexpected within a given context. The idea of singularity of a landscape is therefore related to the surprise produced by the perception of an area in relation to other landscapes previously observed. The observer creates mental expectations based on previous perceptions and experiences, that influence the evaluation process. This 'change of expectation' captures the attention of the spectator and changes his value of the whole. As a result, some landscape elements or objects can act as focal points that increase the value of a certain area, or conversely a spectacular landscape seen prior to another can cause an observer to undervalue the second.

Landmarks or 'singular points' are also an important factor in landscape studies, since they function as unexpected elements in the general perceptive framework of the observed system, or as objects with highly symbolic value that attract attention, becoming structurally defining elements of the space.

Visual fragility can be defined as the 'susceptibility of a given landscape to change, when a new land-use is developed', or 'the deterioration degree of the landscape caused by the incidence of certain activities'. As the definition of landscape units is often linked to artificial features, such as human constructions, sometimes it is easier to assess the degree of 'impact absorption capacity', inverse to fragility, represented by 'the potential of a landscape to absorb or be visually perturbed by human activities' (Ramos 1987).

Finally, 'land use aptness' of an area, refers to its qualities for supporting a determinate land-use with the least possible damage to resources. This concept is of the greatest importance when planning because it serves the purpose of planning while EIA's are for correcting the project.

Given that landscape is a visual resource, visibility (possibility of a point to be seen) and accessibility, are important visual factors, particularly within the framework of environmental impact assessments.

1.2 *Perceptive scales*

We have referred to the specific visual qualities or characteristics of a landscape, but in order to evaluate them it is essential to consider another phenomenon, namely perception. Undoubtedly the act of perception is individual. According to Scott (Blomer & Moore 1982), we can distinguish three distinct magnitudes in a single object:
1. The magnitude that it really has (mechanical measurement).
2. The magnitude that it appears to have (visual measurement).
3. The sensation of magnitude that it produces (sensitivity measurement).

The same distinction could be made from the point of view of colour: the colour that it really is (wavelength); the colour that we can see (depends on the adjacent colours, light intensity, etc.); the symbolic value of this colour (depending on the cultural background of the observer and the symbolism of any given colour in that society). Similar variations could be made with the other visual quality features, and on the synthesis of each unit's characteristics.

Table 2. Different scales of perception of landscape elements

SCALES OF PERCEPTION	TYPE OF ANALYSIS AND UNIT	CHARACTERISTICS
PHYSICAL SCALE	ANALYSIS OF THE UNIT	BELONGING TO THE UNIT: DIMENSIONS, COLOUR TEXTURE, OUTLINE, SHAPE, LINE, ETC
HOW IT IS	UNIT ITSELF	
VISUAL SCALE	ANALYSIS OF THE WHOLE	BELONGING TO THE WHOLE : SHAPE COMPOSITION, SPACES, COLOURS, TEXTURES, LUMINOSITY, SCOPE, ETC.
	RELATIONSHIPS AMONG ADJACENT UNITS	BELONGING TO THE OBSERVER: VISUAL POSITION OF OBSERVER
HOW IT IS SEEN IN RELATION WITH ITS CONTEXT		SINGULARITY, CHANGE IN EXPECTATION, ETC.
SENSIBILITY SCALE	ANALYSIS OF THE WHOLE	BELONGING TO THE OBSERVER: VISUAL POSITION, CULTURE, SYMBOLISM, SINGULARITY, CHANGE IN EXPECTATION ETC.
HOW IT IS EXPERIENCED AND APPRECIATED	RELATIONSHIP BETWEEN QUALITIES OF LANDSCAPE AND QUALITIES OF THE OBSERVER.	

The mere estimation of the individual qualities of each element of the territory is insufficient to develop ranks or scales from this concept. As well as studying how each element 'is' (physical scale), it is necessary to include how it 'is seen' in relation to the surroundings (visual scale) and the feelings of the observer on how it is or how it is appreciated within the cultural context (sensibility scale). The weighted combination of these will give the true value. It could be that an element has a lesser value by itself, but it gives strength to the whole or it has very high symbolic meaning (Table 2).

2 METHODOLOGY

2.1 *Forms of landscape analysis*

The former concepts have been applied to landscape analysis in the Vall de Gallinera (Alicante, Spain), where together with important natural features, human activity has become a significant element of formal and visual value.

Taking into account the special characteristics of the valley, the approach used was based on the consideration of the landscape as a continuous sequence. Three interdependent analyses were applied (Villota 1994) (Table 3).

Table 3. Types of landscape analysis

TYPE OF ANALYSIS	UNITS FOR EVALUATING LANDSCAPE QUALITIES	FACTORS/CHARACTERISTICS USED FOR THE ANALYSIS
1. LANDSCAPE UNITS	EACH ENVIRONMENTAL UNIT AND ITS SPATIAL DISTRIBUTION	INTRINSIC, DESCRIPTIVE AND POTENTIAL VALUES (TABLE 1)
		PHYSICAL CHARACTERISTICS
2. VISUAL BASINS	EACH VISUAL BASIN AS A WHOLE	RELATIONSHIP QUALITIES: CONTRAST BETWEEN ELEMENTS, FORMAL AND CHROMATIC DIVERSITY, ETC.
		VISUAL AND SENSIBILITY CHARACTERISTICS
3. DYNAMIC APPRECIATION ACROSS ITINERARIES	VISUAL SEQUENCES	CHANGES OF EXPECTATION: CONTRAST BETWEEN SEQUENCES, FORMAL STRUCTURAL CHANGES, SINGULAR AND UNEXPECTED ELEMENTS, ETC.
		VISUAL AND SENSITIVE CHARACTERISTICS

The analyses carried out are as follows:

1. The description and spatial distribution of landscape units and types, based on the definition of homogeneous environmental units, is emphasised in order to analyse the precise characteristics of the landscape. Landscape is treated basically from a descriptive point of view, which considers existing features and their intrinsic qualities. Physical scales are used for assesing; however in keeping with the notion that these characteristics cannot be perceptually separated from adjacent units, the influence of visual interactions is also considered for the area's evaluation.

Potential for change is also partially assessed, including intrinsic fragility and capacity to absorb different types of impacts (point, linear or areal) that could be caused by a change of land-uses in the valley. For a complete assessment and mapping of fragility or 'use aptness', it is necessary to know which land uses must be considered in order to establish the types of direct and indirect impacts from each one. A map with all kinds of possible impacts or land uses over each area would lead to confusion.

2. Description of 'visual basins; 'zones from which a point or series of points are "visible'; or 'zones which are visible from a point or series of points' (Claver Farias 1982). The geomorphology of the area, with concave and enclosed land configurations, favours this type of analysis. Here, the valley is treated as a whole system of perceptive relationships in which the visual and sensitive value scales are considered.

3. The dynamic study of the landscape establishes perceptive sequences on the main itineraries; it detects changes in the relative value some spatial units have with respect to neighbouring units, due to the observer's change of expectation. In the case of the Vall de Gallinera, this is included, superficialy, after the visual basins analysis.

The dynamic analysis should be based on the identification of consecutive visual fields, the contrasts between their spatial qualities and the assessment of changing expectations concerning singular elements, structural changes, etc. Visual and sensitive value scales are also considered.

2.2 Landscape analysis using environmental units

The process of landscape mapping using this method starts by the identification of 'geomorphological domains', which give a first classification of the terrain. In the scale in which we work and within the large physiographic units, (watersheds, valleys, depressions, etc) we call 'geomorphological domains' those with similar physiographical position, process (kind of action and agents that determine genesis and similar evolutionary stage.

The synthesis of this map and related features: relief (topography, slope gradient, drainage), geomorphology (landforms, aspect, geology, lithology, structure), leads to the definition of 'Homogeneous units' of the physical environment.

A new combination of homogeneous units and vegetation (wild, cultivated, potential), and anthropic activities (agricultura, industry, settlements, infrastructure, etc.), shows the organisation of the studied area in Environmental Units. According to Cendrero (1975) environmental units are 'territorial units with similar regolithic/soil nature, bedrock, active processes, biological communities and human modifications'. They represent a synthesis of natural and human features

Using the former as a framework and taking into account hydrology (water courses and bodies), geomorphological units and associated hazards (geomorphological units, active processes, hazards), we can interpret the territory from the point of view of: Descriptive visual (Visual quality, Singularity) and Potential features (Fragility, Use aptness).

These parameters are evaluated mainly from the point of view of the physical perceptive scale (Table 2). Finally, the Landscape Map is obtained by interlinking descriptive and potential characteristics of the landscape with the Map of Environmental Units. As a result, the latter are divided into 'landscape units', which represent areas with homogeneous landscape components and visual response to human actions.

The detailed analysis and description of these units was carried out using vertical and oblique aerial photographs but, mainly, by land surveys throughout the territory.

2.3 Landscape analysis using visual basins

As stated above, this method relies on the initial definition of spatial or mapping units, which can be considered as "closed" from the visual point of view. Large-scale geomorphological features, such as crests, valley divides, etc. normally define the boundaries of these basins. The basins can be used for the integrated assessment of descriptive visual characteristics as well as potential ones.

Visual quality, singularity and visual fragility are the parameters which determine the perception and the visual characteristics of each basin. To assess the whole basin 'visual categories', which integrate 'visual quality' and 'singularity', are used. Obviously the assessment is based on the visual and sensitive perceptive scales.

The detailed analysis and description of these kinds of units was also carried out by vertical and oblique aerial photographs that were later used to locate the data obtained by direct surveys.

2.4 Evaluation of visual parameters

For landscape parameters, quality, singularity and fragility 'direct methods of accepted subjectivity' have been applied after discussion among the team members (Claver Farias 1982). A detailed study of the territory following a series of itineraries established with the help of different types of maps and pictures is the basis for discussion. A summarised, although subjective, assessment of the different units was carried out; resulting in a division into categories of a semiquantitative nature, which are easy to understand.

Visual quality of each landscape has been assessed using this last method. The levels established are 'excellent', 'very good', 'good', 'acceptable', 'poor'.

Similarly Singularity has been assessed using simple and easy-to-understand terms: 'very high', 'high', 'medium', 'low', 'very low'.

Landscape categories of the basins, or complex units, are assessed by categories proposed by Fines (1968) (in Claver Farias 1982): 'spectacular', 'superb', 'outstanding', 'pleasant', 'common' and 'ugly'.

Visual fragility is not, strictly speaking, an intrinsic characteristic of the landscape, because it depends on the type of action considered, which will determine the nature and intensity of impacts. In this case, where the actions were not defined, fragility is treated as the other visual characteristics, considering the general vulnerability of units. The value range used was also: 'very high,' 'high', 'medium', 'low', and 'very low'.

In a similar way, but yet more radical, land use capacity depends on the use options considered and an exhaustive study is as laborious as it is useless.

3 THE VALL DE GALLINERA.

La Vall de Gallinera is a municipality of the province of Alicante in the Spanish Mediterranean region, 38°52'-38°46' N and 0°19'-0°10' W, and belongs to the "Comarca de la Marina Alta", located at the northern border of the province (Fig. 1). The Almirante/Albureca and La Foradá ranges mark, respectively, its northern and southern limits. This municipality corresponds approximately to the valley of the Gallinera River (also called Rambla Gallinera), although its boundaries exceed the crests of the above mentioned ranges, and include part of the northern slope of Almirante/Albureca ranges and part of the southern slope of La Foradá range.

3.1 *Geomorphological framework*

The morphology of this valley is determined by its east-west elongation. Steep slopes leave limited open spaces and few plains. Lineal components, such as the river and the cornice of the southern range, characterise the valley and mark the continuity and concatenation of the spaces.
The valley sides are markedly asymmetric in every aspect (topography, geology, morphology, land cover, vegetation, etc.). A geomorphological analysis reveals that the asymmetry of the valley is mainly due to differences in bedrock distribution as well as microclimatic conditions, which favour different kinds of vegetation on each slope.

The cornice of the La Foradá range, an overthrusting front of limestone over marls, tops the southern margin; it forms a marked morphological scarp, which dominates the whole valley. In the middle slope, the regionally called 'Tap marls' crop out. Most human activities are carried out here. The slope face is quite steep and entirely cultivated with tree orchards; retaining walls (locally called 'margens') have been built parallel to the contour lines forming a terraced structure.

The northern margin, composed of limestone, presents a very steep slope face with severe erosive processes taking place, which prevent soil development; conspicuous V-shaped gullies give a characteristic aspect to this side. The lower areas, with lower gradient, are developed on piedmont deposits of limestone, gravel and clays.

The Gallinera River has three distinct geomorphological sectors, which are closely linked with the three visual sub-basins described later (Figs.3 and 4). The upper part, with small intermittent streams, presents a marked V-shaped incision. The central part of the valley is characterised by relatively wide terraces with gravel, sand and silt deposits and poorly developed young soils. A canyon that cuts through massive limestone, produced by a strong fluvial incision due to differential uplift and antecedence processes, dominates the lower, eastern sector.

The southern slope of La Foradá and northern slope of Albureca ranges are formed by karstified limestone, 'lapiaz'. A well-defined poljé forms the northern slope of the Almirante range.

3.2 Analysis of landscape units

The method previously described was applied to the study area, in order to assess its general landscape characteristics and determine potential visual impacts on the different landscape units. Table 4 summarises the descriptive parameters, visual assessment and fragility of landscape units in the study area and Figure 2 represents these units spatially.

Landscape units in the table are initially defined on the basis of broad 'Geomorphological domains' and are denoted by their initial in capital letter: (SU, C, M, B, P, SL, A, V). Each domain is divided according to three different points of view. Physical and morphological features (material, morphology and aspect) are used for the first classification (denoted by numbers 1, 2, 3, etc). A second classification is carried out analysing vegetation (arboreal, scrub and cultivation), indicated by lower case, initial letter (a, s and c). Cases where human activity is relevant are marked with H. It must be taken into account that visual criteria often include different characteristics in one unit; for instance, SL2 contains very different morphologies and slope gradients, but it is a visual unit. Obviously, the number and character of the classes which are established depend on the physical and conceptual scales of the study and, also, on the characteristics of territory and on its diversity.

3.2.1 Description of the Landscape Units
Due to the reduced extension of this paper, only some of the units that present any kind of landscape singularity and only in Gallinera valley will be described; the rest are just listed in the Table of Landscape Units (Table 4).

Geomorphological domain 'Crests'. C-Landscape Units. C1 (s and sH), and C2s. Marked limestone ridges and cornices form this domain, with medium to steep gradients, poorly developed soils and sparse vegetation, mainly scrub; crops are rare. These crests have a marked asymmetry in slope aspect and gradient, which is also reflected in other characteristics. The unit has been divided into three landscape units: C1s and C1sH in northern side, and C2s in southern side.

Landscape unit C2s, with a northern exposure, has very steep slopes, often vertical limestone scarps, on which some rock vegetation has developed. It is present in La Foradá and El Almirante ranges. This prominent east-west unit represents one of the most characteristic features of the Vall de Gallinera. Rock falls and block slides are the main processes and hazards.

Visually, it is a well-defined strip that crowns the whole of the southern slope of the valley. The limestone is massive, grey coloured and textured by frequent vertical lines, conditioned by discontinuities in the limestone. The skyline is complex, with vertical conspicuous elements. Tree-covered slopes mark the lower limit of this unit. In La Foradá, there is marked contrast with the horizontal linear structure of the latter, characterised by terraced fields.

The visual and sensibility magnitude of this unit is enhanced by its dominant position in the valley. Its verticality, set against the sunlight and its contrast with the adjacent units makes this unit seem larger than its real physical magnitude.

The intervisibility relationship with the other points of the valley is very high. This unit is visible from any point, and with a very large solid visual angle.

In accordance with previous considerations, visual quality has been assessed as 'very good' in La Foradá and 'good' in El Almirante. Singularity is 'very high' in both cases.

Considering the dominant physiographic location, the features of the high slope angle, scarce vegetation, active processes, etc. and the high intervisibilty relationship with the whole of the

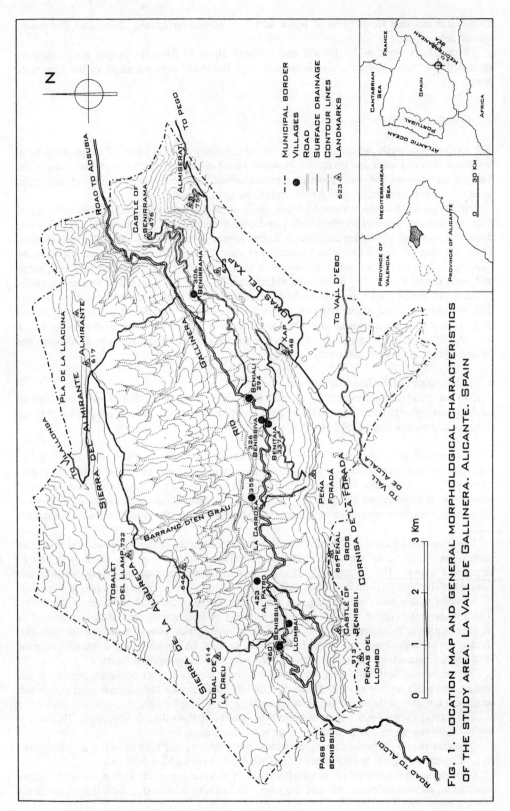

FIG. 1. LOCATION MAP AND GENERAL MORPHOLOGICAL CHARACTERISTICS OF THE STUDY AREA. LA VALL DE GALLINERA. ALICANTE. SPAIN

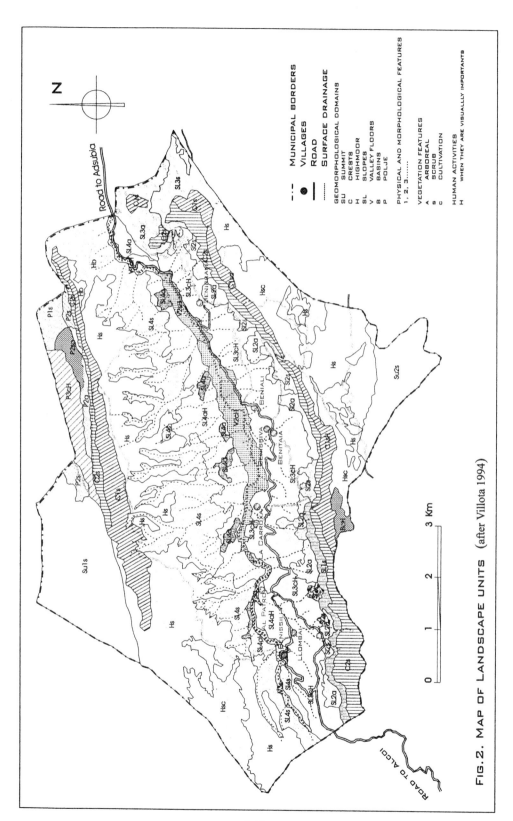

FIG. 2. MAP OF LANDSCAPE UNITS (after Villota 1994)

Table 4. CLASSES OF LANDSCAPE UNITS BY PHYSICAL AND VISUAL PARAMETRES. (modified after Villota 1994)

GEOMOR-PHOLOGICAL DOMAIN	LANDSCAPE UNITS CLASSIFICATION ACCORDING WITH:			PHYSICAL AND MORPHOLOGICAL FEATURES (1)				VEGETATION FEATURES (2)						HUMAN ACTIVITIES (3)		
	(1)	(2)	(3)	GEOLOGY MATERIALS	GEOMORPHOLOGY MORPHOLOGIES	MAIN	ASPECT	ARBOREAL (a) DENS.	DISTRIBUTION	SCRUB (s) DENS.	DISTRIBUTION	CULTIVATION (c) DENS.	DISTRIBUTION	BUILDINGS & ESTRUCTURES	EXTRACTIONS
SUMMIT SU	SU1	SU1a		DOLOSTONES	DOLINES V-SHAPED VALLEYS	VAR.	20-40%	RARE	SPARSE	MED.	UNIFORM	RARE	ISOLATED POINTS		
	SU2	SU2a		LIMESTONES	V-SHAPED VALLEYS	VAR.	20-60%								
CRESTS C	C1	C1a		LIMESTONES	LAPIAZ	S	20-40%	LOW	DISPERSED	MED.	UNIFORM	RARE	ISOLATED POINTS	DISPERSED STOCKYARDS	CLAYS
		C1aa						MED.	UNIFORM			NULL			
	C2	C2a		LIMESTONES	WALLS & CORNICES	N	>40%	RARE	SPARSE	LOW	DISPERSED CLUMPS	NULL			
HIGMOOR H	H	Ha		LIMESTONES	DOLINES MAMELONS LAPIAZ	S	5-20 %	LOW	DISPERSED	HIGH	UNIFORM	NULL LOW	ISOLATED POINTS	INNER PATHS	
		Hac													
SLOPES SL	SL1	SL1a		DOLOSTONES	REGULATED-SLOPE,	N	20-40%	RARE	SPARSE	HIGH	UNIFORM	NULL			GRAVELS
	SL2	SL2a		COMGLOMERATES	COLLUVIUM	N	20-40%	LOW	DISPERSED	HIGH	UNIFORM	NULL		TANKS FOR IRRIGATING	
		SL2a						HIGH	UNIFORM	MED.	UNIFORM	RARE	ISOLATED POINTS		
	SL3	SL3ac		MARLS	U-SHAPED VALLEYS LOW-RIDGES	N	20-40%	HIGH	UNIFORM	MED.	UNIFORM	MED.	UNIFORM	BENCH-TERRACES	
		SL3c	SL3cH				10-20%	MED.	DISPERSED CLUMPS	LOW	DISPERSED CLUMPS	VERY HIGH	UNIFORM	BENCH-TERRACES ROAD BENCH-TERRACES VILLAGES	
					LEDGES	N	<10%					HIGH	DISPERSED CLUMPS		
	SL4	SL4a		LIMESTONES	V-SHAPED VALLEYS PIEDEMONT	S	40-80%	RARE	SPARSE	MED.	UNIFORM	RARE	ISOLATED POINTS	BENCH-TERRACES DISPERSED STOCKYARDS	GRAVELS
		SL4aa	SL4aaH	AGLOMERATES		S	10-40%	MED.	DISPERSED CLUMPS						
VALLEY FLOOR V	V1	V1a	V1aH	LIMESTONES CONGLOMERATES	CANYON	VAR.	>40%	HIGH	UNIFORM	MED.	UNIFORM	RARE	ISOLATED POINTS	ROAD	LIMESTONES
	V2	V2c	V2cH	ALLUVIAL AGLOMERATES	TERRACES FLOOD PLAIN PIEDMONT	E	<10%	LOW	DISPERSED CLUMPS	LOW	DISPERSED CLUMPS	VERY HIGH	UNIFORM	BENCH-TERRACES	GRAVELS
		V2a	V2aH					RARE	SPARSE	MED.	UNIFORM	MED.	UNIFORM	MILLS	
	V3	V3a		LIMESTONES	V-SHAPED VALLEYS	VAR.	20-40%	RARE	SPARSE	MED.	UNIFORM	NULL			
		V3a						LOW	DISPERSED CLUMPS	HIGH	UNIFORM	RARE	ISOLATED POINTS		TAP FLAGSTONES
BASINS B	B	Bac	BaacH	MARLS	LOW RIDGES	VAR.	10%	MED.	DISPERSED CLUMPS	MED.	DISPERSED CLUMPS	MED.	DISPERSED CLUMPS	BENCH-TERRACES	
KARSTIC DEPRESSION POLJE P	P1	P1a		DOLOSTONES		W	20-40%	RARE	SPARSE	MED.	UNIFORM	NULL			
	P2	P2a		AGLOMERATES	ALLUVIAL	E	20-40%	LOW	DISPERSED	MED.	UNIFORM	LOW	DISPERSED CLUMPS	BENCH-TERRACES	
		P2aa				N	20-40%	HIGH	UNIFORM						
						N	20-40 %	MED.	UNIFORM						
	P3	P3c	P3cH	DECALCIFIED-CLAYS	DECALCIFIED-CLAYS	VAR.	<5%	HIGH	UNIFORM	HIGH	UNIFORM	MED.	UNIFORM	SETTLEMENT OF ISOLATED HOUSES	

Table 4. CLASSES OF LANDSCAPE UNITS BY PHYSICAL AND VISUAL PARAMETRES. (modified after Villota 1994)

GEOMORPHOLOGICAL DOMAIN	LANDSCAPE UNITS CLASSIFICATION ACCORDING WITH:			PROCESSES/ HAZARDS		SINGULAR POINTS		VISUAL CHARACTERISTICS				VISUAL FRAGILITY		
	(1)	(2)	(3)	TYPE	DEGREE	ARCHITECTONICS	GEOMOPHOLOGIC	VISUAL QUALITY	SINGULARITY	LANDSCAPE CATEGORY	POINT	TYPE OF IMPACT		AREAL
												LINEAR		
SUMMIT SU	SU1	SU1a		INCISION EROSION	LOW			GOOD	MEDIUM	PLEASENT	MEDIUM	HIGH	HIGH	HIGH
	SU2	SU2a		KARSTIFICATION	HIGH			GOOD	MEDIUM		MEDIUM	HIGH	HIGH	HIGH
CRESTS C	C1	C1a		HYDRIC-EROSION	MED.			GOOD	HIGH	PLEASENT	MEDIUM	MEDIUM	MEDIUM	MEDIUM
	C1sa							GOOD	HIGH		MEDIUM	MEDIUM	MEDIUM	MEDIUM
	C2	C2a		BLOCK-ROCK FALLS & TOPPLING	MEDIUM	BENIRRAMA CASTLE	PENYA FORADA PENYAL GROOS	VERY GOOD	VERY HIGH	SPECTACULAR	VERY HIGH	VERY HIGH	VERY HIGH	VERY HIGH
HIGMOOR H	H	Ha		KARSTIFICATION FLUVIAL INCISION	HIGH HIGH			ACCEPTABLE	MEDIUM	PLEASENT	MEDIUM	MEDIUM	MEDIUM	MEDIUM
		Hac		HYDRIC EROSION	MEDIUM			ACCEPTABLE	MEDIUM		MEDIUM	MEDIUM	MEDIUM	MEDIUM
SLOPES SL	SL1	SL1a		INCISION	HIGH	BENISSILI CASTLE		VERY GOOD	MEDIUM	SUPERB	MEDIUM	HIGH	HIGH	HIGH
	SL2	SL2a		INCISION HYDRIC EROSION	MEDIUM			GOOD	HIGH		MEDIUM	HIGH	VERY HIGH	VERY HIGH
		SL2a									MEDIUM	MEDIUM	VERY HIGH	VERY HIGH
	SL3	SL3abc		LANDSLIDE CREEPING	HIGH			GOOD	HIGH	SUPERB	MEDIUM	MEDIUM	VERY HIGH	VERY HIGH
		SL3c	SL3cH	HYDRIC EROSION SWAMPING	MEDIUM VERY LOW	BELL TOWERS		VERY GOOD	VERY HIGH		VERY HIGH	VERY HIGH	VERY HIGH	VERY HIGH
	SL4	SL4a		INCISION HYDRIC EROSION	VERY HIGH		BARRANCO DEL GRAO	VERY GOOD VERY GOOD	MEDIUM MEDIUM	OUTSTANDING	HIGH HIGH	VERY HIGH HIGH	VERY HIGH	VERY HIGH
		SL4aa	SL4aaH											
VALLEY FLOOR V	V1	V1a	V1aH	INCISION BLOCK-ROCK FALLS	MEDIUM LOW		KARSTIC CANYON	GOOD	HIGH	OUTSTANDING	VERY HIGH	VERY HIGH	VERY HIGH	VERY HIGH
	V2	V2c	V2cH	FLUVIAL EROSION FLOODING	LOW MEDIUM			VERY GOOD	MEDIUM	PLEASENT	MEDIUM	MEDIUM	MEDIUM	MEDIUM
		V2a	V2aH								HIGH	MEDIUM	MEDIUM	MEDIUM
	V3	V3a		INCISION	HIGH		EXHUMED FORDA FALL PLAIN	ACCEPTABLE	LOW	PLEASENT	VERY HIGH	VERY HIGH	VERY HIGH	VERY HIGH
		V3a						VERY GOOD	HIGH		HIGH	HIGH	HIGH	HIGH
BASINS B	B	Bac	BacH	HYDRIC EROSION CREEPING	MEDIUM			ACCEPTABLE	LOW	COMMON	LOW	LOW	LOW	LOW
KARSTIC DEPRESSION POLJE P	P1	P1a			LOW			ACCEPTABLE	MEDIUM		MEDIUM	HIGH	HIGH	HIGH
	P2	P2a		HYDRIC EROSION KARSTIFICATION	MEDIUM		AS A WELL-CONSERVED WHOLE	GOOD	HIGH	OUTSTANDING	MEDIUM	MEDIUM	MEDIUM	MEDIUM
		P2aa									HIGH	MEDIUM	HIGH	HIGH
	P3	P3c	P3cH	KARSTIFICATION SWAMPING	HIGH	PLEASURE URBANIZATION		GOOD	HIGH		MEDIUM	MEDIUM	MEDIUM	MEDIUM

valley, visual fragility is 'very high' for any kind of impact. For the same reasons, and especially due to active processes, it has very low aptitude to support any change in land-use.

Geomorphological domain 'Slopes' SL. Landscape units SL1s, SL2a, SL3sac, SL3cH, SL4s, SL4saH. This is the most widely represented unit in the study area, occupying most of the Gallinera valley along both river margins, and constitutes the closure of the valley. It is heterogeneous, due to the asymmetry of the two valley sides. Geomorphologic features (see paragraph 3.1), solar radiation conditions (the northern slope receives an excess of solar radiation, which makes it unsuitable for agriculture, while the southern one, on the shady side, is more suitable for different crops), vegetation, human activities and visual contrast, mark the differences.

Along the valley, two strips clearly visible from the right slope, are in the shade. The narrower one (SL2 unit), on the top of the slope and under the crest, is formed by conglomerates of colluvium with a little development of soils, with abundant trees, partly autochthonous. The units on marl (SL3 units) situated beneath the above mentioned one, form a wide strip where most human activities are developed. Slope gradient is quite steep; soils are poorly develop and subject to erosion. Flat-bottomed streams cross the unit.

The left slope (SL4 units), on the sunny side, presents limy materials and steep slopes with marked erosional processes that prevent soil development; it also has V-shaped gullies. The lower part of the slope, with a lower gradient, forms a piedmont deposit of limestone, gravels and clays.The upper part of the slope is covered by scrub vegetation while the lower area also has scattered clumps or individual trees.

Landscape unit SL3cH occupies most of the right slope and is an important unit for the configuration of landscape. Most human activities are concentrated here; it is completely terraced and covered by tree crops.

The main geomorphologic processes are creeping, sliding and rill erosion; retaining walls of terraces mitigate their effects. Rock falls and flash floods are the main hazards.

Visually it is a well-defined unit limited by the river on the lower side and, on the upper one, by the green strip of SL2 unit crowned by the Foradá cornice with its vertical structure. The 'margens', formed by limestone, run parallel to the contour lines, giving a staircase pattern to the landscape, with a horizontal component that underlines the linear character of the valley. The white colour of the soils together with the grey colours of the limy 'margens' and the different shades of green of crops, provides a complex colouring pattern which contrasts with the uniformity of the other units, enhancing its own uniqueness. Open spaces are scarce, formed by structural ledges which favour small dispersed settlements. Some of the villages are built halfway along the slope, and act as landmarks, which reinforced by the bell towers of their churches, highlight the linear structure of the valley.

The intervisibilty relationship with other units is very high. This unit is visible from any point of the valley, always with a large solid visual angle; the other units are also clearly visible from this point.

The unit on marls, with its great diversity, chromatic and structural complexity, has 'very good' visual quality. Slope landforms and the contrast with adjacent units, increase the singularity of this unit that can be assessed as 'high'. Intrinsic visual fragility against point or horizontally linear impacts is low, while it is very high to any extensive or linear impacts that break the staircase pattern of the unit.

Geomorphological domain "Valley floor". V. Landscape Units V1aH, V2a and V2s. This domain, the valley floor of the Rambla Gallinera, can be divided, from the geomorphological and the landscape points of view, into three different sectors. (See 3.1)

1. Landscape unit V1. Source and upper part of the Gallinera River. It is a marked gully where hydric erosion prevents the development of soils. Low scrub vegetation and oleanders on the gully floor; occasionally re-forested with pine trees.

Visually it is a narrow, deep space, ochre and grey with green areas. It is a characteristic riverside landscape in karstic zones. Intrinsic visual quality is 'good', and singularity 'medium'.

Intervisibility relation with the others units is very low, and it is far from the most frequented paths. Consequently, although its own fragility is high, it could be assessed as 'medium' in relation to the other units.

2. Landscape unit V2. Alluvial plain of Rambla Gallinera. The plain is completely cultivated. The most significant process are flash floods.

Visually it is the only flat area of the valley; well cared for crops give it a tidy look. It presents a grained, green-coloured and uniform texture Visual quality is 'acceptable' and singularity 'low'. Intervisibility with other units is very high and it is easily accessible. Visual fragility is thus assessed as 'medium', but it is very sensitive to impacts from large structures.

3. Landscape unit V3. Formed by a lineal and sinuous karstic canyon. Abundant tree and scrub vegetation in the canyon area, with crops on certain parts of its floor. Intervisibilty with the other units is very low; however, visibility and accessibility are very high because the main road follows it.

Visually it is a narrow area, between close-vertical walls, ochre and grey with dry-green areas. It is a very characteristic landscape of karstic zones. Intrinsic visual quality is 'very good", singularity is "very high" and it has been considered as a 'singular area' because of its intrinsic interest as well as the role it plays in the whole of the valley. Intrinsic visual fragility is 'high' for every type of impact. However, the road offers some small strips where linear impacts can be absorbed.

3.3 Visual basins analysis

Three visual basins and three sub-basins have been defined in the study area (Fig. 3). The former are clearly bounded by the divides of the Vall de Gallinera (Almirante/Albureca and La Foradá ranges) and they are the visual areas of Pla de la Llacuna (north), Vall de Alcalá (south) and Vall de Gallinera itself (centre). The latter is divided into three sub-basins: Al Patró (upper valley), Benissiva (middle valley) and Benirrama (lower valley). The limits between these sub-basins are not well defined, but they show clear differences in spatial qualities, shape, texture, ground cover, land use and structure. It is interesting to note that the three sub-basins roughly coincide with the three parishes established in the seventeenth century (Al Patró, Benissiva and Benirrama).

3.3.1 The southern visual area (Vall de Alcalá)
It forms part of the Vall de Alcalá visual basin and it mainly consists of a "highland" furrowed by several deep gullies presenting lapiaz-like erosion. Scrub with dispersed trees are the dominant land cover. Landscape qualities of this unit should be assessed refering to other areas not described here. A first evaluation could be "common".

3.3.2. El Pla de la Llacuna (The Laguna Plain)
This visual basin is formed by a large poljé whose southern part belongs to the municipality of Gallinera. It is a karstic depression that forms a typical poljé, whose development has been favoured by the existing system of faults. Diversity of its landscape units is very high from all points of view; the terrain goes from flat to more than 40% gradient slopes, from "terras rossas" to dolostones, vegetation ranges from scarce to very high density, etc.

Vegetation, except in an area reforested with pines, consists of medium density scrubs and a few dispersed trees. The poljé floor is bound by colluvium along its southern border, with abundant scrub-like vegetation and some areas reforested with trees. It has medium grained texture, dun and a dun-green colour sprinkled with dots of coarse-grained texture determined by tree cover. The rest of the depression is almost flat and covered by "terra rossa", occupied by crops and mainly by a sparse settlement.

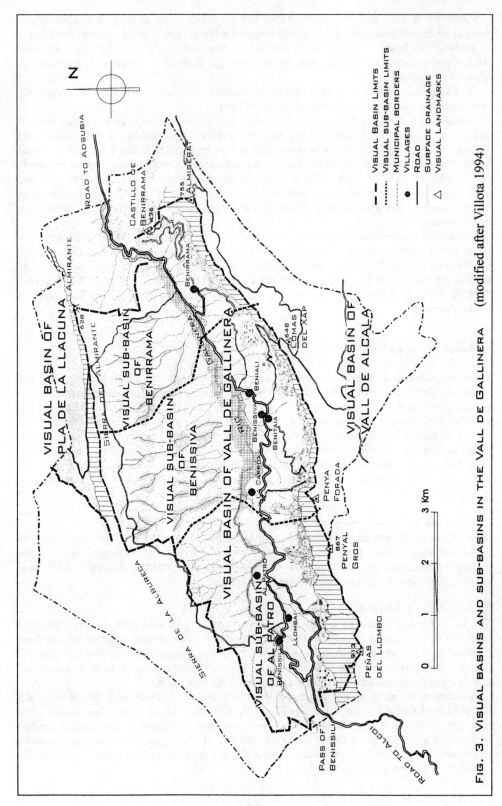

FIG. 3. VISUAL BASINS AND SUB-BASINS IN THE VALL DE GALLINERA   (modified after Villota 1994)

Visually, it forms a plain surrounded by strong escarpments with contrasting colours, a attractive unit with certain cultural value due to its geological singularity (the well preserved poljé). An 'outstanding' landscape category has been given to this unit due to its singularity, cultural value and geologic features. Consequently, the area presents high visual fragility except on the floor of the poljé, where it is 'medium'.

3.3.3 *The Gallinera Valley*

The valley forms a whole visual basin clearly visible from the upper part of the slope. It presents a clear asymmetry; the northern slope (sunny side) forms a surface with low scrub vegetation, medium grain textures and grey-dun colours; a colluvium strip with some trees links this slope to the valley floor. Deep V-shaped gullies cross the slope. The southern side (shady margin) has been previously described (paragraph 1 in 3.2.1.3.) It is a complex unit made up of a succession of shapes, volumes, colours and textures, parallel to each other and to the bottom of the valley with increasingly steep slopes that form a staircase structure.

Visually, the Gallinera basin is formed by a series of interconnecting spaces which together form a whole, whose formal, textural and chromatic richness is increased by the asymmetry of its slopes, as well as by the contrast of luminosity between them due their east-west trend aspect. Many landmarks close to the crests and also on half-slope enhance the landscape and help to structure it.

Three sub-basins can be identified within the valley (Fig. 4), although they present similar lateral closures they have different spatial structure and specific characteristics.

Sub-basin of Benirrama includes in the eastern part of the area the karstic canyon previously described. Its western limit can be established west of Benirrama where the valley becomes wider. Visual characteristics are determined by the contrast between the steep rock sides of the canyon and the valley floor dominated by the green and reddish colours of oleanders.

Sub-basin of Benissivá includes the central and widest part of the valley, whose western limit can be defined around La Carroxa. This area is characterised by the relatively wide open spaces formed by the alluvial plain and terraces covered by orange trees, which produce a clear contrast with the ochre and green colours of scrub-covered northern slope and the rain-fed crops on the southern one. Its configuration, width, clear order and harmony make it the most important sub-basin of the whole area.

Sub-basin of Al Patro, to the west, includes the V-shaped part of the valley; erosion processes are intense here and prevent the development of vegetation, which is limited to sparse scrub and some pines planted on terraced land. Tributary streams into the main river are also markedly incised and they form spaces of limited visibility.

Landscape category for the Gallinera valley basin and its sub-basins has been considered as 'superb', for its singularity, cultural value, geomorphological features and formal richness. Although very fragile in general, some of its landscape units can absorb the impact of medium-size structures.

3.4. *Dynamic analysis*

The dynamic analysis considers the fact that the viewer is not static, but moving through the landscape, Coming from the east, the karstic canyon previously described, is a linear, closed, narrow and sinuous one, in constant shade due to the presence of abundant trees and the vertical cut of the canyon itself. The relative darkness thus created gives a sense of visual adaptation between the open spaces of the Adsubia area and the more confined valley, creating an atrium effect that enhances the openness and luminosity of the areas further downstream. From the western side of the valley, looking from the top of Benissili pass, the valley appears to be a completely closed space with the landmark of the castle of Benirrama, in the middle of the visual

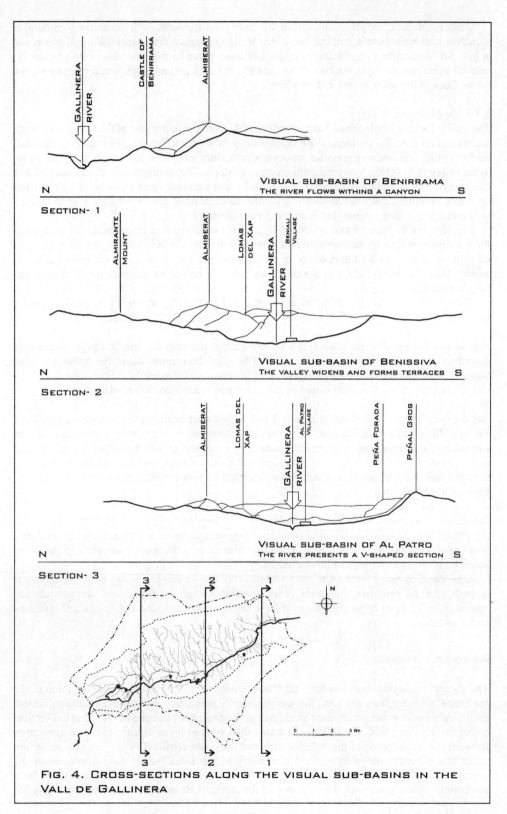

FIG. 4. CROSS-SECTIONS ALONG THE VISUAL SUB-BASINS IN THE VALL DE GALLINERA

angle and standing out against the sky, which increases visual value by a "surprise effect" or change of expectation.

3.5 *Characteristics of landscape and singular points*

The general characteristics of landscape in the study area can be summarised as follows:

1. Geological and geomorphological features are an important conditioning factor for visual perception in the area; The Crests in the northern slope and the deep gullies in the southern one (especially the Barranc d'en Grau, in front of La Carroxa, that almost cuts the hillside) enhance the asymetry.

2. Human elements (bench terraces, villages) are an essential feature of the landscape and they reinforce the visual qualities of the natural space.

3. Human activities have been smoothly integrated within the environment, creating a new type of landscape with harmonic contrasts and a high aesthetic value.

4. On the whole, the area has very good landscape quality and high fragility. This means that it is very sensitive to visual impacts and that future activities (if significantly different from present ones) could affect it negatively.

There are several singular geological and architectural structures that stand out as visual singularities and act as focal landscape elements. (Fig. 5).

1. Peña Foradá: A natural arch excavated into the crest of de La Foradá range, clearly visible from almost every point of the valley.

2. Block slides: Large limestone blocks detached from southern scarps and which have slid the valley, constituting prominent landscape features. Particularly outstanding is the block of Benirrama, on which the castle is built.

3. Church towers. They stand out as landmarks along the valley reinforcing its lineal structure.

4. Benissili and Benirrama castles built on elevations close to the crests, they also stand out and are clearly visible from most of the study area.

5. Exhumed fault plane near Benissili. Very attractive geomorphological feature on the western limit of the valley, uncovered by differential erosion of the contact between the limestone and the Tap marls. Its value is mainly cultural and scientific rather than visual.

3.6. *Sinthesis of results*

The results obtained with the analysis of landscape units and visual basins together with the dynamic analysis are presented in Figure 6. For easier comparison, the concept "visual category" (Table 4), which combines "visual quality" and "singularity", is used in both static analyses. In the dynamic analysis, the main elements and units that structure the landscape are indicated and also assessed in similar terms.

Visual basins –large scale units - are useful to assess possible aptness for future actions as well their visual impacts from a general point of view. The dynamic analysis shows the most important elements from the point of view of landscape structure; these are elements that determine internal relationships which help to understand landscape.

Landscape units -small scale units- are elelements that facilitate specific land-use planning, through mapping and assessment of aptness and impacts.

The sinthesis of the results obtained following the different analyses has an interpretative character and depends on the final objectives of the assessment: land use aptness or EIA.

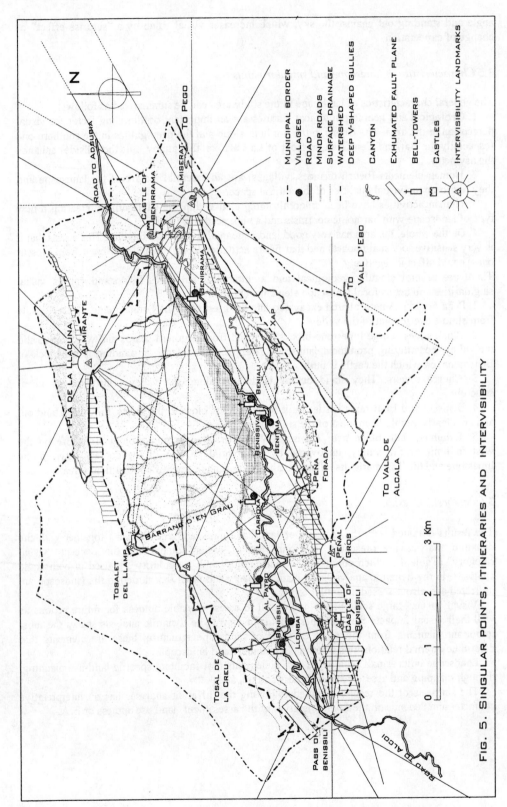

FIG. 5. SINGULAR POINTS, ITINERARIES AND INTERVISIBILITY

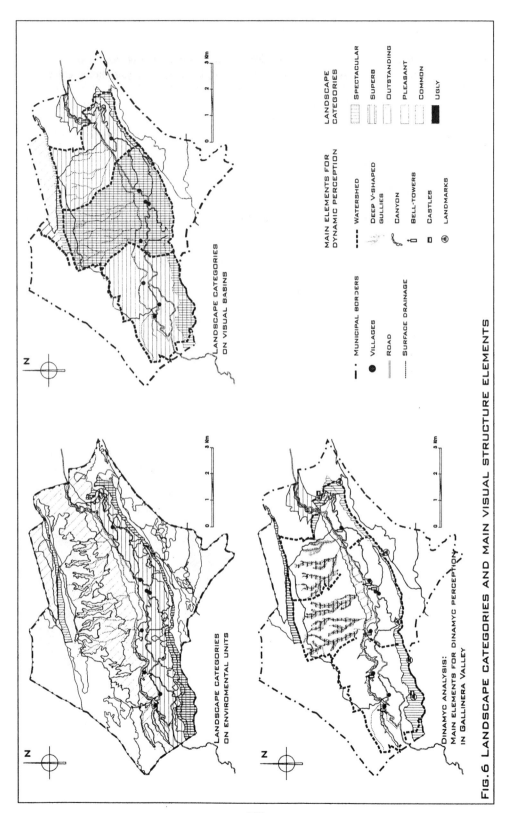

FIG.6 LANDSCAPE CATEGORIES AND MAIN VISUAL STRUCTURE ELEMENTS

4 CONCLUSIONS

This analysis is based on the premise that landscape perception cannot be considered as a static characteristic but rather as a dynamic process. Therefore, landscape analysis should emcompass both static (landscape units and visual basins) and dynamic (visual paths) dimensions. These three different ways of analysing landscape are complementary; each one of them qualifies the others.

The nature of landscape assessment includes different scales of perception: physical (how it is), visual (how it is seen) and sensitive (how it is felt). The appreciation of a landscape image should thus include, at least at some stage, the consideration of wide irregular units (environmental units or visual basins) in which a given element is not only perceived by itself, but also in relation to surrounding ones and to other visual characteristics.

Geomorphological features represent a very adequate basis for the identification and mapping of environmental units which can be used for landscape assessment. Large scale geomorphological features are also essential for the definition of visual basins which provide a general framework for landscape analysis.

The use of geomorphological domains and environmental units has the advantage of providing a natural basis for the integration of the most significant and permanent landscape characteristics: the configuration of the space and processes affecting its evolution. On the other hand, these units do not always coincide with actual perceptual units; landscape value, which often results from the relationship between adjacent units which are conceptually separated using this approach, may not be properly assessed in that case.

The study of the environment and its image, the landscape, can be performed by the sucessive integration of its different characteristics (geomorphologic, biologic and anthropic) in homogeneous units of different levels of synthesis. This system, based on the definition of wide regular units, has the advantage over regular geometric units of better showing the spatial relationships between a point and its surrounding features.

"Natural" environmental units also provide the means to express the transformation experienced by the environment due to natural causes and changes in land use. They can thus serve to assess evolutionary processes of the landscape.

Although landscape can be described in terms of its individual elements or components, the mere analysis of these components is not sufficient for landscape assessment, especially at certain scales. Visual assessments must consider landscape as a whole system, in which the different elements are integrated and inter-related in such a way that any alteration in one of them affects the characteristics of the others. Synthetic, subjective assessments are particularly adequate here. Accepted subjectivity methods, when applied by experts, provide an easy, systematic evaluation of an area, but have the drawback of not allowing comparisons with other areas.

Be it through the use of environmental units or visual basins, the assessments which combine the description of physical elements and visual characteristics provide the basis for identifying the parts of a study area with the highest quality and fragility and therefore most sensitive to impacts from human activities.

The method proposed provides the means to carry out both general assessments for land use planning and more detailed, for EIA os specific actions.

ACKNOWLEDGEMENTS

This work has been supported by the following research projects: DGICYT PB95-946, Junta de Castilla-León SA64-94, and 1FD. 97-222.

REFERENCES

Bloomer, K.C. & Moore C.W. 1982. *Cuerpo, memoria y arquitectura. Introducción al diseño arquitectónico*. Madrid. H. Blume ediciones

Cendrero, A. 1975. *El mapa geoambiental en la evaluación de recursos naturales y en la planificación del territorio*. Santander. Secc. Pub. de la Universidad de Santander.

Claver Farias, I. (1982). *Guía para la elaboración de estudios del Medio Físico: Contenido y metodología (Medio Físico, Análisis y tratamiento de la información)*. Madrid. M.O.P.U. Serie Manuales.

Montes, C. 1992. *Representación y Análisis Formal. Lecciones de Análisis Formal*. Valladolid Manuales y textos universitarios. Secretariado de Publicaciones. Universidad de Valladolid.

Ramos, A. 1987. *Diccionario de la Naturaleza. Hombre, ecología y paisaje*. Espasa Calpe, Madrid

Villota, I. de. 1994 "*Estudio de arquitectura popular en la zona alta septentrional de Alicante y su integración en el Medio Ambiente: La Vall de Gallinera*. PhD Thesis. Escuela Técnica Superior de Arquitectura de Madrid. Universidad Politécnica de Madrid. Unpublished.

Villota, I. de. et al. 1995. La Vall de Gallinera; Un Recurso Geológico-Paisajístico. *Monografia sobre a la I Reunión Nacional de la Comisión de Patrimonio Geológico*.108-115. Madrid. Ed. Comisión del Patrimonio Geológico (Sociedad Geológica de España). Fundación Gómez Pardo.

Villota I de, et al. 1996. Análisis de paisaje por evalución de cuencas visuales y unidades ambientales: La Vall de Gallinera. Alicante. España. *Sexto Congreso Nacional y Conferencia Internacional de Geología Ambiental y Ordenación del Territorio: Riesgos naturales, Organización del Territorio y Medio Ambiente*. Granada.

The use of models for the assessment of the impact of land use on runoff production in the Ouveze catchment (France)

Th.W.J.Van Asch & S.J.E. Van Dijck
Department of Physical Geography, Utrecht

ABSTRACT: In this case study an example will be given on how models and measuring experiments can be used to estimate the effects of land treatment and land use change on runoff production. Model simulations were carried out on slopes with sandy loam soils in the Ouveze catchment in France. Simulation with a physically based runoff infiltration model including lateral flow through macro pores and the soil matrix, may give some important information about the impact of different factors on the production of runoff on these slope. A stepwise procedure is used where in a systematic way the runoff of a 450 m long slope without a channel system is compared with the runoff production of equally long slopes but varying in slope angle, surface topography related to different tillage operations, hydraulic conductivity in depth and vegetation. Concentration of water into ploughing furrows and wheel tracks is an important factor in runoff production on the agriculural slopes. Also compaction, due to the wheel burden of tractors, in the subsoil increases the amount of runoff production. So according to the model outcome, deeper ploughing and flattening of the slope surface would be important remedial measures to mitigate the amount of surface runoff. Shallow soils in the areas with shrub and forest vegetation do not function as a buffer for runoff production during high magnitude storm events. Macro pore flow is predicted not to contribute significantly to the storm hydrograph.

1 INTRODUCTION

Within the procedures of Environmental Impact Assessment it is a difficult task to assess the rate of changes of geomorphological and environmental processes due to human interference.

The problems which arise may be caused by a lack of knowledge of the operating systems in the landscape, the lack of data available from the region and the short time which is available to make planning decisions (Brunsden 1995). To address these problems geomorphologists make use of so called geomorphological indicators which give some estimates of intensity, frequency and magnitudes of processes which took place in the past. For example for flooding there are enough indicators to estimate the magnitude and frequency of floods and peak discharges in the river bed. However these indicators do not give information on the relative contribution of the different source areas in the catchment to the magnitude of discharge in the channel. Therefore the impact of land use changes or remedial measures to mitigate peak discharges and flooding is difficult to assess. No clear geomorphological indicators are available to evaluate the runoff response for different land units in a catchment. One can think of developing a rating system giving scores to landscape factors influencing the runoff response in different source areas. This factorial scoring can be based on general experiences e.g. how soil and vegetation factors react on infiltration and runoff. Morgan (1995) mentioned several problems in using this scoring technique which are related to the classification of landscape factors, the interaction between factors and the problem how to treat the factorial scores in order to come to a general score.

A severe problem is to define the range of the scale for each individual landscape factor, which determines the weight, which is given to each factor. In order to get some idea of the range of the scale of different factors, models in combination with field measuring can be used to estimate these ranges. In this case study, an example will be given on how models and measuring experiments can be used to estimate the effects of land treatment and land use change on runoff production on slopes.

2 THE IMPACT OF LAND USE ON RUNOFF PRODUCTION IN SOME LAND UNITS IN THE OUVEZE CATCHMENT

The area which is selected for this investigation belongs to the 35 km^2 catchment of the Groseau, an affluent of the Ouveze. The catchment forms part of a oligocenic synclinal, filled up with Miocene deposits. The deposits consists of bioclastic and sandy limestones, blue marls calcareous sandstones and marly limestones. Most agricultural land is found on soils developed in marl, unconsolidated sandstones and in recent colluvium. Former rangelands covered by bush and shrubs are found on soils developed in the Jurassic or Cretaceous calcareous substrate. These soils are thin and have high stone contents. Forests are found on both substrates. The soils developed in marl, unconsolidated sandstone and colluvium are used for fruit culture and viniculture. These cultures require mechanical and chemical treatment of the soil.

To control water competing weeds, the soil between the crop rows is kept bare in soil strips of about 6 meter between the fruit trees and 2.5 meter between the vineyards. The soil between the fruit trees and vine grapes is tilled to a depth of 10 cm with a harrow. This is done to remove the weeds, to fight rats and mice and to increase the porosity of the soil.

Before the start of the winter the soil is tilled to 20 cm depth with a plow to allow rain which fall in the next spring to be stored in the soil. Plowing may create temporary macropores where the capillary force is nil.

For the tillage operation and chemical treatments (from march to september once in a month and in may june two times in a month) tractors are used. The wheel tracks left by the tractors are about 30 cm wide. Due to the frequent loading of the tractor, the soil is compacted below the plowing horizon at a depth of about 15 cm (Van Dijck 1996).

2.1 *The model*

Since a couple of years the Department of Physical Geography in Utrecht has developed a GIS which comprises a Dynamic Modelling module, a Cartographic Modelling module and a Geostatistical Modelling module, which are connected to PCraster. The Dynamic Modelling module is integrated at a high level with the part of the package for GIS functions and the Cartographic Modelling module. It provides a meta-language within which the user can build a dynamic model with the operators that are also used for Cartographic Modelling. Extra operators are added for creation of iterations through time and the reading of time series.

The Dynamic Modelling language can be used for building a wide range of models from very simple (point) models up to conceptual complicated or physically based models. For this study a physically based model was developed to simulate the effect of different types of land use and treatments on the production of runoff. The model describes the infiltration into the soil matrix and macro pores, percolation of water through a stack of discrete layers, lateral moisture distribution within either the soil matrix system or the macro pore system, or from macro pores into the soil matrix and vice versa, return flow from the macro pores and overland flow. The concept was developed by Bronstert (1995).

Several other simulations have been carried out to analyse the effect of land use on runoff production in the Groseau area by Koole & Veltman (1996) using the LISEM model and Amaru et al. (1995), using GISFLOW. These model applications showed interesting results on a catchment scale but they did not provide a systematic overview of the effect of separate landscape factors on runoff production. The authors used the one layer and two layer Green and Ampt model, while in this study the different hydrologic conductivities for eight layers and the presence of macro pores is introduced. The effects of the presence of a tillage layer, soil compaction, development of macro pores by plowing and the root formation under natural vegetation, were simulated by adjusting values for the concerning parameters in the model for a run on a 45 m slope in a hypothetical agricultural field of similar length, hereafter referred as "standard slope".

Values for the model parameters of this slope are summarized in Table 1.

Table 1. Values of the model parameters for the standard slope.

length	45 m
width	10 m
slope angle	3%
soil texture	sandy loam
Ks	1.1×10^{-5} m / s (saturated hydraulic conductivity)
Macro porosity	0
Manning's	n = 0.05
van Genuchten *	a = 0.005, m = 0.66, n = 1.2
surface state	smooth, unrilled, unchanneled

* *Van Genuchten (1980): Parameters for calculating the unsaturated hydraulic conductivity from soil water retention curves*

The standard slope is assumed to represent bare soil strips between the crops or fruit yards, aligned in the direction of the slope along a length of 45 m. On the slope a standard rain storm was applied with an intensity of 50 mm per hour and a duration of one hour.

2.2 *The simulation of different impacts on the land*

Figure 1 shows the cumulative total amount of surface runoff in m^3 / m as a function of time during the model rain storm. The effect of plowing and compaction of the soil on the standard slope is simulated here. Plowing of the soil means the development of macro pores in the soil which in this simulation is estimated to be 5% of the soil volume. Measurements in the
field have revealed that after plowing the porosity may decrease by 4%, which must be ascribed to consolidation of the soil (van Dijck 1996) and the collapse of the macro pores.

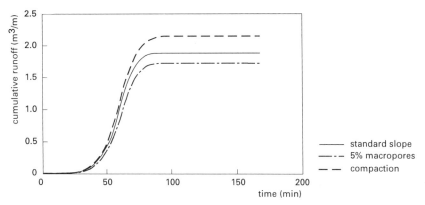

Figure 1. Cumulative discharge of water from a standard slope in agricultural bare land with a sandy loam soil compared with the same slopes with macro pores (due to plowing) and compaction of the subsoil from a depth of 15 cm through tractor loading.

Figure 1 shows that the effect of macro pores may reduce the amount of runoff production on this standard slope by 9%. Rainfall simulation tests in the field however showed no significant differences in infiltration capacity of a freshly plowed soil and soils which were not plowed during a month (van Dijck 1996). Obviously the formation of macropores during plowing is overestimated for these soils with sandy loam texture. Measurements revealed that due to compaction, the Ks values of the subsoil at a depth of 15 to 20 cm show a decrease by 50% in the saturated hydraulic conductivity. This was simulated in another run shown in Figure 1. The model produces a significant increase in runoff production by 17% compared with the standard slope. Up till now rainfall experiments have been carried out which show less dramatic effect of a compacted subsoil on the infiltration capacity.

Figure 2 shows the effect of macro relief on the runoff production in agricultural areas. The effects of 2 types of macro relief were simulated separately: plowing of the soil making 2 furrows per meter of 10 cm width and 5cm depth and the formation of wheel tracks of 30 cm width and 2 cm depth.

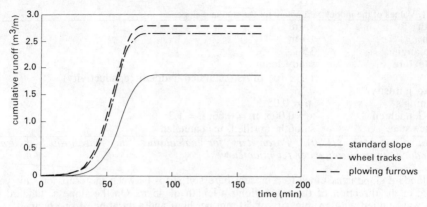

Figure 2. Cumulative discharge of water from a standard slope with a flat surface in agricultural bare land with a sandy loam soil compared with the same slopes with wheeltracks (30 cm width, 2cm depth and plowing furrows 10 cm width and 5 cm depth.

The effect on the production of runoff is high: an increase in total runoff production of respectively 49% and 44% for plowing furrows and wheel tracks. The output of the model shows (not depicted here) that the concentration of water into gullies increases the runoff velocity and hence reduces the residence time of water on the slope, reducing the total amount of infiltration. The concentration of water in linear elements (furrows, wheeltrack ditches and from the parcels into other linear elements like paths, roads and small drainage canals is a well know feature on slopes in agricultural catchments. The linear concentration of runoff water in these agricultural catchments is an important factor which is often neglected in runoff studies Papy & Boiffin (1989).

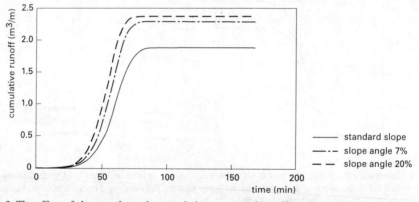

Figure 3. The effect of slope angle on the cumulative amount of runoff production.

The same effect of an increase in runoff velocity, which could be observed in the model output, occurs when the slope angle increases. Figure 3 shows the effect of an increase in slope angle to 7% (increase in runoff production 22%) and an increase of the slope towards 20% (increase in runoff 28%). It is remarkable that a change in slope gradient from 3% to 7% has more effect than a change from 7% to 20 % (Fig. 3).

Figure 4 gives some information on the effect of forest and shrub vegetation on runoff production. On the more or less natural slopes which in the area are steeper (20% in the simulations), the runoff production is quite different: forest or shrub slopes with deep sandy soils and a high amount of macro pores (in the simulation 15%), give the lowest runoff production (decrease of runoff with respect to standard slope 39%, see curve "natural/vegetation deep soil" in Figure 4). However in these areas many natural slopes have a thin soil, not more than 15 cm in depth, with high amount of macro pores but with a rather impermeable rock beneath (2.7×10^{-7} m / s in the simulation). Runoff production for these slopes was also simulated (see Figure 4 "natural vegetation shallow soil") and the model produces a runoff amount, around 2 m^3 which has the same order of magnitude as in the agricultural fields, despite the high amount of macro pores. One has to keep in mind however, that the effect of the more or less impermeable subsoil will become apparent for long term and/or high intensity rain events.

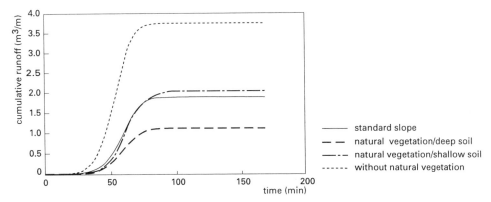

Figure 4. The effect of natural vegetation and the removal of natural vegetation on the amount of cumulative runoff compared with the standard slope in soils with a deep sandy loam profile and shallow soils with a nearly impermeable rock substratum on slopes with an angle of 20%. The removal of vegetation is done on the slope with shallow soils.

Figure 4 also shows the effect of the loss of natural vegetation, which e.g may occur through fires, on these shallow thin soils. It was assumed that in that case the amount of macro pores decreases from 15 to 5% and the hydrological conductivity from 15 to 8 cm per hour. The effect on runoff production can in that case be dramatic as is shown in Figure 4.

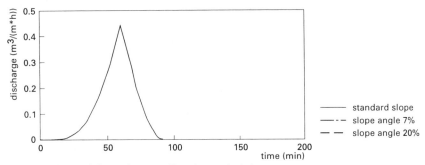

Figure 5. The hydrograph for surface runoff on the standard slope.

Figures 5, 6, 7 show the contribution of macro pore flow to the hydrograph for soils in agricultural land with and without compaction (5% volume of macro pores) and soils in forested areas (15% volume of macro pores) in thin and deep sandy loam soils. First one can conclude that the contribution of macropore flow (Figs. 6, 7) to the runoff hydrograph (Fig. 5) is very small, 0.2% at most. Further one can deduce from these Figures that the volume of pores and the slope angle has a great influence on the amount of macro pore flow. Compare e.g. in Figure 6 the soil in agricultural land with no compaction (pore space 5%) with the deep soils in the forest (pore space 15%). Also the effect of a less impermeable sublayer can be observed. Especially in the forest areas with a nearly impermeable substratum (Fig. 7), the amount of macro pore flow increases with a factor around 600, compared with the deep forest soils (Fig. 6). The second flat peak in both figures in the hydrograph is caused by the influx of water from the nearly saturated matrix into the macro pores.

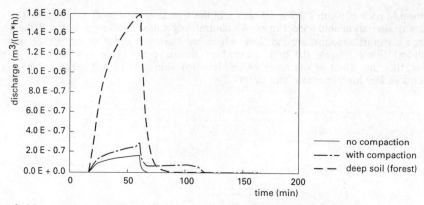

Figure 6. Macro pore flow hydrograph for a slope (slope angle 3%) in agricultural bare soils with compaction and no compaction in the subsurface (volume of macropores is 5%) and for a slope (slope angle 20%) with natural vegetation and deep soils with a macro pore volume of 15%.

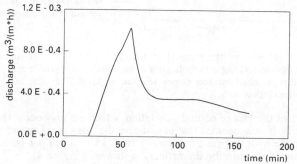

Figure 7. Macro pore flow hydrograph for a shallow soil with a nearly impermeable substratum under natural vegetation (slope angle 20%) and a macro pore volume of 15%.

3 DISCUSSION AND CONCLUSIONS

The use of hydrological models in combination with field measurements is an essential tool for an environmental impact assessment of land use and tillage on runoff production. Despite the fact that there is always a lack of data in such procedures and that the models do not give an accurate prediction of the real situation the use of these models may give some feeling of the relative weight of the impact of different factors in the system. A pure qualitative approach based on experience knowledge can lead to serious errors in estimating the impact effects for different land use scenarios. Sensitive analyses with models may provide a better understanding of the system and can be used as a support for the setting up of a certain type of rating system for the area.

A first analysis of different types of land use and tillage operations in the Ouveze catchment (Groseau sub catchment) on slopes with sandy loam soils, gave some surprising results of some impact factors, which could have been wrongly estimated in case of a pure qualitative approach. The results, based on model simulations are not always verified by measurements and therefore may be wrong. But the use of this model makes us aware of the potential importance of different impact factors.

The here presented model simulations, with input of some hydrological parameters of the sandy loam soils, showed a relative important effect on runoff production of tillage operations on the land: the creation of furrows by plowing and the formation of wheel tracks give a considerable amount of water concentration which increases the total amount of runoff production during non steady state conditions. Therefore an important reduction of runoff can be achieved when after each tillage operation the soil is made more or less flat. Also the compaction of the soil can be considered as a factor increasing runoff production. Plowing increases the amount of macro pores but the effect on runoff reduction in the output of the model is not so high as might be expected. Moreover field measurements learned that there was no effect of plowing on the infiltration capacity of the soil. As might be expected reforestation on sandy loam soils with a deep profile, considerably decreases the amount of runoff production. It appeared also that runoff production initiated by large storm events on shallow soils in forested or shrub areas is quite high and comparable with runoff production in the agricultural areas despite the high macro porosity of the soil. Forest fires therefore produces areas with potentially high run-off production on these shallow sandy loam soils. The contribution of macro pore flow to the storm hydrograph is rather limited for agricultural areas as well as for more or less natural areas.

REFERENCES

Amaru, M.,Van Dijck, S.J.E. & Van Asch, Th..J. 1995. Predicting the impact of land use change on event-based runoff in a small Mediterranean catchment. . In M. Panizza, A.G. Fabbri, M. Marchetti & A. Patrono (eds), Special EU Project Issue: Geomorphology and Environmental Impact Assessment. *ITC Journ.* 1995 4: 331-336.

Bronstert, A. 1995. User manual for the Hillflow-3D catchment modelling system (Physically based and distributed modelling of runoff generation and soil moisture dynamics for micro-catchments). Working Document 95/4, Cooperative Research Centre for Catchment Hydrology. Karlsruhe: Institute for Hydrology and Water Resources Planning, University of Karlsruhe.

Brunsden, D. 1995. EIA and geomorphologic processes. In M. Panizza, A.G. Fabbri, M. Marchetti & A. Patrono (eds), Special EU Project Issue: Geomorphology and Environmental Impact Assessment. *ITC Journ.* 1995 4: 339.

Koole, D. & Veltman, S 1996. Afvoermodellering in het stroomgebied van de Groseau (Vaucluse, Frankrijk). Utrecht: Graduate report, Department of Physical Geography, University of Utrecht.

Morgan, R.P.C. 1995. Soil Erosion and Conservation. Essex:Longman.

Papy, F. & Boiffin, J. 1989. The use of farming systems for the control of runoff and erosion . In U. Schertmann, R.J. Rickson & K. Auerwald (eds), *Soil Technology* Series 1: 29-38.

Van Dijck, S.J.E. 1996. The influence of land use on water and sediment transport from a cultivated catchment in the Provence (France) during single rain storms. Internal report, Utrecht: Department of Physical Geography, University of Utrecht.

Van Genuchten, M.Th. 1980. A closed form equation for predicting the hydraulic conductivity of unsaturated soils. *Soil Sci. Soc. Am. J.*, 44: 892–898.

Floods in the city of Rio Branco, Brazil: A case study on the impacts of human activities on flood dynamics and effects

Edgardo M. Latrubesse
Universidade Federal de Goiás-IESA, Campus II, 74001-970, Goiânia-GO, Brazil

Samia Aquino da Silva
Universidade Estadual de Maringá-UEM, Geography Department, 87020-900, Maringá, PR, Brazil

Maria de Jesus Moraes
Universidade Federal do Acre, BR 364 km4, Campus, 69915-900, Rio Branco, AC, Brazil

ABSTRACT: Floods of the Acre River river were analysed through hydrological and geomorphological data. Floods affect thousands of people along the alluvial plain/lower terrace of the Rio Acre in the city of Rio Branco, capital of the State of Acre. During the latest flood of 1997, approximately 70,000 people were affected. The case study illustrates some of the consequences of ill-planned urban growth and provides some ideas for better incorporating the consideration of geomorphological processes into EIA.

1 INTRODUCTION

One of the most frequent types of human impacts on geomorphological processes are the ones derived from urban expansion and related human activities. In the case of the Amazon region, urban growth normally takes place in an indiscriminated manner on river banks which have been previously cleared.

Population growth in the Amazonia during the last decades has concentrated in the main Brazilian cities. The case of Rio Branco, capital city of the state of Acre, Brazil, serves to illustrate this problem. The population of Rio Branco has exploded during the last two decades, growing from 89,993 inhabitants in the 80's to 167,442 in the 90's. The consequent land occupancy took place in a chaotic manner, without appropriate urban planning nor any evaluation of environmental impacts. As a result, thousands of families settled on the alluvial plain of the Rio Acre, which crosses through the city and divides it into two parts. These families are regularly affected by floods and mass movements which produce large economic losses and social problems.

In this paper we focus on one of the impacts derived from the process of urban growth described. The flood dynamics of the Rio Acre and its consequences for the city of Rio Branco, the most important city in the state of Acre and the Acre fluvial basin (Fig. 1) are analysed and assessed. This case study illustrates this kind of environmental problems, common to several cities along the Amazon basin.

2 METHODS

The work presented is here based on the analysis of stage and discharge data at Rio Branco and Brasileia gauging stations, available from the ANEEL (National Agency of Electric Power). Geomorphological mapping was carried out using aerial photographs at the scales 1:40,000 and

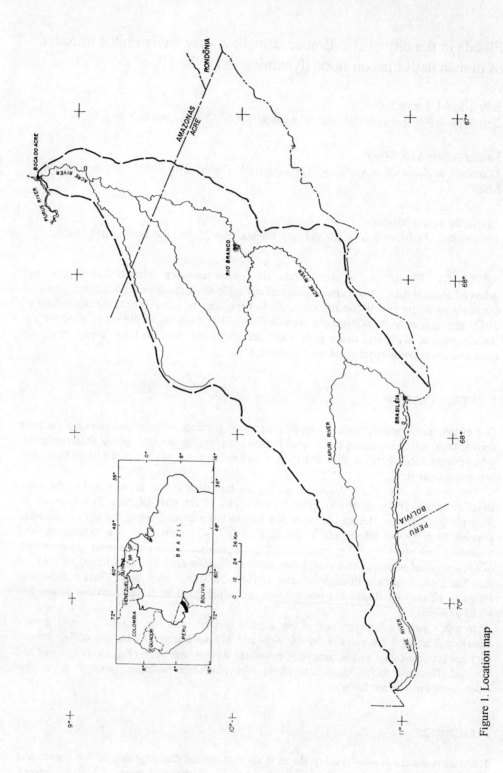

Figure 1. Location map

1:8,000, as well as topographic maps at scales 1:10,000 and 1:2,000. Population data were obtained from different public agencies. Field work was carried out between 1994 and 1997.

3 THE ACRE RIVER

The Rio Acre is a tributary of the Purus River; its basin is developed on Tertiary claystones, silstones and fine sandstones of the Solimões Formation (Upper Miocene-Pliocene). The climate is tropical and the area is covered by tropical rainforest. The headwaters of the basin are at an altitude below 350 m, in Peruvian territory. The Acre is a typical river of the South-western Brazilian Amazonia, with a single and sinuous channel and asymmetric and complex meanders alternating with straight segments. These types of tropical rivers in the South-western Amazonian lowlands could carry approximately 98% of their sedimentary load in suspension (Gibs 1967, Mertes 1985).

The alluvial belt of the river is formed by three terrace levels. The older terraces reach 15-30 m and 34-38 m above the lower water level of the river. The younger terrace is 8-12 m high. In Rio Branco, this lower terrace is formed by a complex meander belt with meanders at different stages of formation (Figure 2; compare with urban area in Figure 9).

The older unit (stage 4) occupies the area of the International Airport on the right side of the river. It is a flat area with some meanders totally filled. The intermediate unit (stage 3) has paleomeanders easily identifiable on aerial photographs. They occupy a flat area on the right side

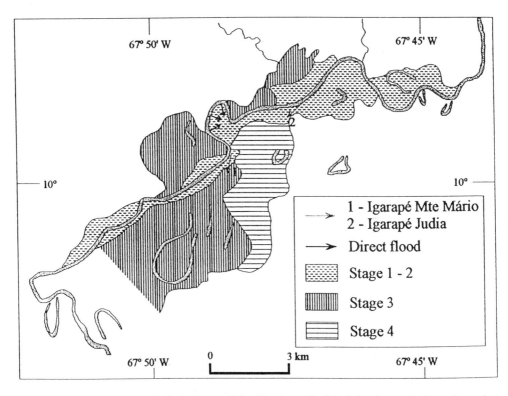

Figure 2. Simplified geomorphological map of the Rio Acre alluvial plain. Arrows indicate the main flood direction along Monte Mario and Judia Igarapes.

of the river between the city and Amapa Lake. The more conspicuous landforms here are large filled meanders, up to 130 m wide. These meanders can be identified by their morphology, hindered drainage and swamp vegetation.

The modern unit (stages 2 and 1) is located along the channel, on both sides. The width of this unit ranges between 0.4 and 2.4 km. Some partly filled, abandoned meanders can be recognised. The meanders and abandoned channels are 70-80 m in width, have poor drained soils and swamp vegetation.

The presently active channel is nearly 75 m wide. In Rio Branco it forms a large meander which cuts into the Tertiary sediments of the Solimões Formation, cropping out at the lower part of the banks.

4 HYDROLOGY

Records of stage and daily discharge in the Rio Acre, obtained at the Rio Branco gauging station from 1971 to 1993, and complementary data from the Brasileia gauging station in the upper Acre River, were analysed. Stage data for 1997 were also considered. The Rio Branco gauging station is located in the city of Rio Branco, where the drainage area of the Rio Acre is 22,670 km^2. At Brasileia, the drainage area is 3,299 km^2

The mean annual discharge in Rio Branco during the period analysed was 350 m^3/s. The Rio Acre has clear seasonal peaks of high and low discharge (Fig. 3). High flows are concentrated between January and April; some secondary peaks also occur during this period. The system has a pronounced annual variability between high and low discharges. The ratio between minimum and maximum average discharge for a day reaches 72.9. The average of maxima is 1463 m^3/s

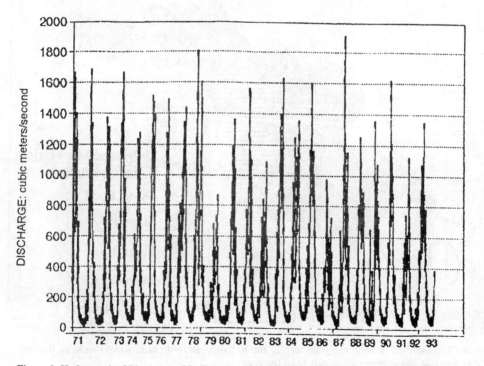

Figure 3. Hydrograph of Rio Acre at Rio Branco gauging station, 1971-1993; mean daily discharges.

Table 1. Annual highest and lowest mean daily discharges and ratio between maxima and minima; Rio Branco gauging station, 1971-1993.

Year	Maximum		Minimum		Ratio
	Date	Discharge	Date	Discharge	
1971	Feb. 24	1677	Sept. 20	23.0	72.9
1972	March 01	1688	Aug. 18	27.2	61.3
1973	Feb. 23	1411	Sept. 03	30.8	45.8
1974	March 04	1779	Aug. 20	30.5	58.3
1975	March 22	1276	Oct. 31	39.6	32.2
1976	Feb. 17	1508	Oct. 17	43.2	34.9
1977	Feb. 27	1484	Sept. 18	29.0	51.2
1978	Dec. 26	1786	Sept. 05	46.5	38.4
1979	March 29	1691	Sept. 18	47.5	35.6
1980	March 26	850	Sept. 17	33.2	25.6
1981	March 26	1353	Oct. 09	32.0	42.3
1982	Feb. 26	1553	Sept. 08.13	44.0	32.3
1983	March 23	1201	Oct. 02. 03	25.7	46.7
1984	April 12	1654	Oct. 07	52.0	31.8
1985	April 28	1456	Aug. 17	64.8	22.4
1986	Feb. 09, 10	1623	Oct. 05	41.7	38.9
1987	Jan. 19	1006	Aug. 03	20.4	49.3
1988	Feb. 17	1914	Nov. 03	46.0	41.6
1989	Feb. 13	1316	Oct. 07	37.6	35.0
1990	Jan. 06	1340	Aug. 31	38.4	35.0
1991	Jan. 25	1616	Oct. 17	31.2	51.8
1992	March 27	1151	Sept. 05	36.0	32.0
1993	March 11	1332	Sept. 24	39.6	33.6
Average		1462.8		35.5	41.2

Table 2. Annual highest and lowest daily stages (m); Rio Branco gauging station, 1971-1993.

Year	Highest		Lowest		Range	Ratio
	Date	Stage	Date	Stage		
1971	Feb. 24	16.30	Sept. 20	2.80	13.50	5.8
1972	March 01	16.25	Aug. 18	2.94	13.31	5.5
1973	Feb. 23	14.80	Sept. 03	3.06	11.74	4.8
1974	March 04	16.86	Aug. 20	3.05	13.81	5.5
1975	March 22	14.00	Oct. 31	3.29	10.71	4.3
1976	Feb. 17	15.36	Oct. 17	3.38	11.98	4.5
1977	Feb. 27	15.23	Sept. 18	3.00	12.23	5.1
1978	Dec. 26	16.90	Sept. 05	3.45	13.45	4.9
1979	March 29	16.38	Sept. 18	3.47	12.91	4.7
1980	March 26	11.38	Sept. 17	3.13	8.25	3.6
1981	March 26	14.46	Oct. 09	3.10	11.36	4.7
1982	Feb. 26	15.61	Sept. 08,13	3.40	12.21	4.6
1983	March 23	13.55	Oct. 02	2.89	10.66	4.7
1984	April 12	16.17	Oct. 07	3.56	12.61	4.5
1985	April 28	14.87	Aug. 17	3.76	11.11	4.0
1986	Feb. 09, 10	15.72	Oct. 05	3.26	12.46	4.8
1987	Jan. 19	12.34	Aug. 02,03	2.68	9.66	4.6
1988	Feb. 17	17.11	Nov. 03	3.15	13.96	5.4
1989	Feb. 13	14.18	Oct. 07	2.94	11.24	4.8
1990	Jan. 06	14.34	Aug. 31	2.95	11.39	4.9
1991	Jan. 25	15.82	Oct. 17	2.78	13.04	5.7
1992	March 27	13.32	Sept. 05	2.89	10.43	4.6
1993	March 11	14.28	Sept, 24	2.98	11.30	4.8
Average		15.01		3.13	11.88	4.8

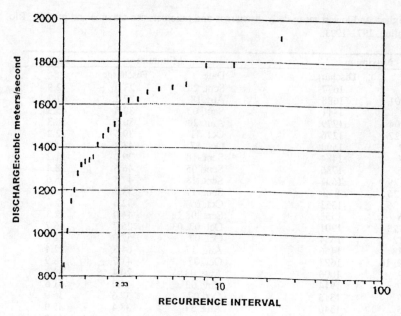

Figure 4. Recurrence interval of floods at Rio Branco for the period 1971-1993. Maximun daily discharges for each year were used.

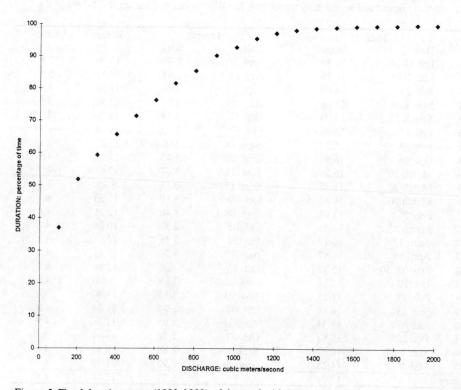

Figure 5. Flood duration curve (1983-1993), elaborated with mean daily discharges for the entire period.

and the average of minima is 35.5 m^3/s (Table 1). The range in stage is also large (Table 2). The average difference in water surface elevation between high and low stages is approximately 8.75 m.

The recurrence interval of floods was determined for a period of 23 years. The interval used to determine the mean flood was 2.33 years (Leopold et al., 1964). The Acre shows a low flood variability from year to year (Fig. 4). The flood of 1988, which reached 1914 m^3/s (the largest value recorded), is only 1.3 times the discharge of the average annual flood.

A 13.5 m stage for bankfull is used by the Civil Defence as the alert stage for floods. This bankfull stage/discharge is close to the value of bankfull discharge with a recurrence interval of 1.5 years in the annual flood series.

The flow duration curve was elaborated with the daily discharges from 1983 to 1993 (Fig. 5). Floods have short duration in comparison with the long period during which low discharges are characteristic. Over 60 % of the time the river has a discharge equal or less than the mean annual discharge of 350 m^3/s, and bankfull discharges are present less than 5% of the time.

The channel of the Acre is deep and narrow. The widht/depth ratio was determined by relating bankfull width and mean depth and it is relatively low: 8.8.

5 CLIMATE AND DISCHARGE

Southwestern Amazonia has a humid tropical climate. Mean annual rainfall is nearly 2000 mm. However, Southwestern Amazonia has a pronounced dry season between June and September. This variability in precipitation is related to the shift of the ITCZ (Inter-tropical Convergence Zone). During the Summer of the southern hemisphere the ITCZ is located between 10° and 15° S. Southern Amazonia receives most of its rainfall during this period. The ITCZ shifts northwards, reaching its extreme north position at Venezuela and Colombia, from July to August. During this period, rainfall drops significantly in South-western Amazonia, which has a well-defined "dry season" during Winter.

Rainfall records from the Rio Branco station at the Federal University of Acre, for 1971-1994 are shown in Table 3. The average value from June to August is approximately 115 mm, a very low value for a tropical humid area which supports a rainforest.

Mean monthly discharge and precipitation are plotted in Figure 6. The period represented includes a year of with a high flood (1988) and two "normal" years (1989-1990). There is a general tendency to a linear response: increasing precipitation-increasing discharge and decreasing precipitation-decreasing discharge. However, the behaviour and runoff response to precipitation might be more complex than initially expected. We do not know the mechanisms of storage in the flood plain from one year to the next or during the same year. As stated by Nordin and Meade (1982) for the Amazon basin, "the highest stage of the year should not be expected to correlate with total annual precipitation or even with seasonal or monthly point values as much as with the temporal and spatial distribution over the drainage basin".

The tropical humid climate of Amazonia is characterised by strong convective showers; amount and spatial distribution of precipitation can be highly variable within short distances. This can be inferred from Figure 7, which shows the values of monthly discharges at the Brasileia and Rio Branco gauging stations. As a first approximation, we could say that discharge behaviour in both stations is coincident. However, a "random" behaviour is detected in Figure 7.

In spite of the scarcity of data, some tendencies can be drawn from the climate-discharge relationship. First, the high and low discharge peaks coincide roughly with the periods of high and low precipitation (rainy and dry seasons). Second, the river response to increased discharge during the rising stages occurs nearly two months after the start of the rainy season. This response is very different during falling stages . The abrupt decrease in precipitation during the beginning of the dry season is accompanied by a strong drop in discharge/stage values. Third, differential peaks (known in Brazil as "repiquetes") of high discharge which are out of phase

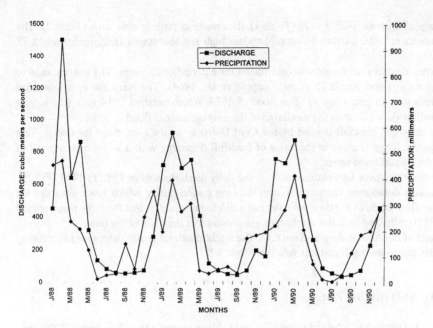

Figure 6. Monthly precipitation and discharge at Rio Branco gauging station (1988-1990)

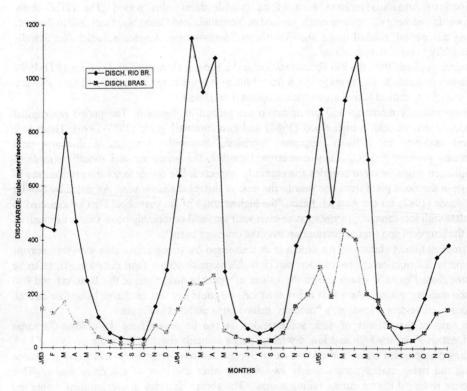

Figure 7. Monthly discharge at Rio Branco and Brasileia gauging stations (1983-1985)

Table 3. Monthly rainfall (mm), standard deviation, mean and rainfall variability at Rio Branco from 1971 to 1994.

Year	Jan.	Febr.	Mar	April	May	June	July	Aug.	Sept.	Oct.	Nov.	Dec.
1971	157	246	292	210	139	16	18	80	86	214	191	275
1972	260	258	167	183	123	21	33	49	23	136	275	155
1973	250	271	247	162	23	48	68	63	138	55	328	207
1974	391	250	138	132	123	90	10	28	68	94	114	199
1975	210	352	171	151	54	79	49	51	137	121	210	274
1976	402	296	88	204	54	7	0	3	18	209	252	260
1977	251	279	272	53	91	81	153	17	110	295	201	275
1978	211	257	165	228	130	0.6	124	19	103	181	242	318
1979	214	177	365	96	176	1	5	109	19	208	94	252
1980	203	429	237	44	88	18	47	30	159	134	99	230
1981	265	218	139	155	32	1	5	74	138	251	171	209
1982	376	359	182	158	193	38	45	21	97	202	196	202
1983	269	241	292	159	57	20	50	14	55	132	197	288
1984	374	352	261	169	99	4	18	10	154	228	201	180
1985	160	141	244	381	44	28	114	98	80	167	283	369
1986	286	357	343	188	235	10	27	85	98	258	249	284
1987	361	180	137	180	24	25	90	20	48	154	285	287
1988	448	465	233	204	124	13	28	33	147	52	250	347
1989	194	392	272	303	46	35	51	62	38	169	181	194
1990	219	279	411	205	74	14	6	31	116	185	200	282
1991	273	215	184	103	125	21	12	2	236	104	187	268
1992	198	388	325	168	105	33	49	36	116	86	92	226
1993	451	337	300	302	47	89	26	13	61	122	244	137
1994	206	403	248	185	43	54	1	57	63	159	115	227
Std.Dv..	89.516	85.225	80.979	74.713	55.883	28.268	41.062	30.620	52.725	63.166	64.805	57.136
Mean	276.21	297.58	238.04	180.13	93.708	31.108	42.875	41.875	96.167	163.17	202.38	247.71
Variab.	0.32	0.28	0.34	0.41	0.59	0.86	0.95	0.73	0.54	0.38	0.32	0.23

between the Rio Branco and Brasileia stations, reflect water inputs from tributaries (Japuri River and others). This could be indicative of differences in precipitation within short distances, which can differentially affect the sub-basins and produce "repiquetes".

6 THE FLOODS

Floods affect the city of Rio Branco every year. The most important floods properly recorded were those of 1988 and 1997. During the flood of 1988 the river reached a stage of 17.11 m and the maximum daily discharge was approximately 1914 m^3/s. During the flood of 1997 the stage reached was 17.66 m and the discharge was nearly 2,008 m^3/s.

The dynamics of a flood in this river does not correspond to generalised overbank flow. During annual floods, overbank flow affects a small part of the lower terrace. Similar behavior was observed during the large flood of 1997. The main damages were produced by reflux of water into the lower terrace through smalls tributary channels named Igarapé Monte Mario and Igarapé Judia. The Monte Mario Igarapé (regional name for small tributaries) occupies in its lower course a paleochannel of the Acre River on the lower terrace. The presence of a narrow natural levee on the banks of the Acre River impedes direct flooding over a large portion of the lower terrace. Flood waters penetrate upstream from the Acre River to the Igarapé Monte Mario and Judia, flooding stages 1, 2 and part of stage 3 of the lower terrace (Fig. 2). According to available records, the higher stage 4 has not been affected by floods.

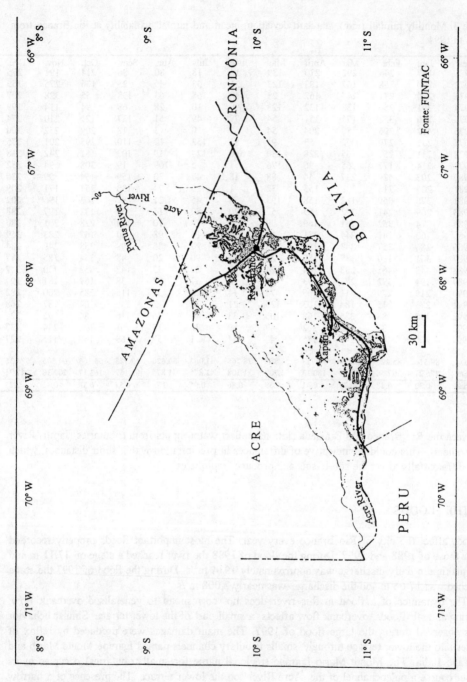

Figure 8.- Deforested area (dotted) of the Rio Acre fluvial basin. Deforestation is concentrated along state highway BR 317.

7 FLOODS AND VEGETATION

It has been estimated that at least 50% of annual rainfall in these regions is the result of evapotranspiration from the rainforest (Salati et al., 1978). The rainforest plays another important role from the hydrological viewpoint. Rainforests store water and can sustain evapotranspiration when the air temperature increases and moisture is reduced. Deforestation for the implantation of cattle ranches replaces the rainforest by pastures, which are poor water storers (Walker 1991). Consequently, this results in an increase in water runoff and a decrease in water storage. However, it is not easy to establish a clear relationship between deforestation and flooding or deforestation and river responses in areas with incomplete hydrological and rainfall records. Therefore, it is difficult to know to what extent peak stages are a reflection of runoff. It is also difficult to know how much water is stored in the floodplain and the effect of this water on discharge. In these tropical rivers vegetation and floodplain play the role of interactive "sponges" which can regulate or affect fluvial dynamics.

Forest loss is quite important in the Rio Acre basin; Figure 8 shows areas affected by deforestation (FUNTAC, 1989). Deforestation occurred as a result of the establishment of "fazendas" or large cattle ranches, located mainly along state highway BR 317. Nearly 13 % of the basin experienced rapid deforestation during the last two decades. Although we cannot drive quantitative conclusions concerning to what extent deforestation affected the fluvial system, it is reasonable to consider, in view of the analysis above, that decreasing or reversing deforestation in the basin will reduce flood impacts in the future.

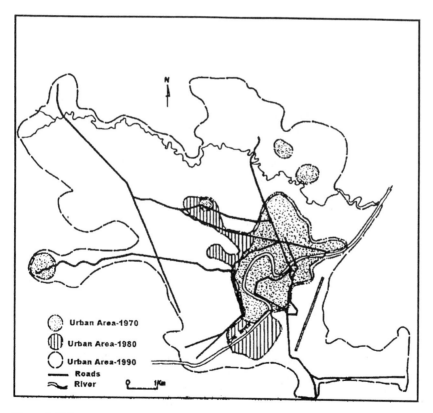

Figure 9. Urban expansion of Rio Branco, 1970-1990.

8 FLOODS AND PLANNING

The expansion of cattle ranches, the abandonment of "seringais" (native rubber stands) and the arrival of immigrants from the Mid-West and South of Brazil produced a period of rapid population growth, especially since the seventies. In 1970 the city had approximately 20 neighbourhoods and 35,500 inhabitants. In 1990 it grew to 150 neighbourhoods and 167,442 inhabitants (IBGE, 1991), with a significant increase of the urban area (Fig. 9). This accelerated urban expansion, combined with a lack of urban planning, resulted in settlements in areas of low slope gradients, headwaters of drainage systems ("igarapés") and along the floodplain and lower terrace of the Rio Acre -in periodically or occasionally flooded areas. This occurred despite the recommendation by the City Planning Directory (Prefeitura Municipal 1986) that these areas should not be occupied.

During the flood of 1988, 23 neighbourhoods along the fluvial belt of the Rio Acre, and Igarapés Judia, monte Mario and São Francisco were affected; it is estimated that approximately 20,000 persons were affected to a greater or lesser extent. During the flood of 1997 the figures were 41 neighbourhoods and approximately 70,000 people. However, in 1997 the discharge was only about 5% greater and water stage 0.49 m higher than in 1988; the area of the lower terrace affected by flooding was only slightly larger than during the 1988 flood.

This illustrates the types of impacts which could be considered in (and avoided through) the process of urban planning. The 1988 flood event provided a good example of the impacts to be expected from fluvial dynamics in the basin. Failure to consider this when planning the expansion of the city resulted in much greater damages in 1997; the latter could have been easily avoided or reduced using simple geomorphological knowledge and tools (Fig. 2) for planning the new settlements.

9 FINAL REMARKS

The case study presented here illustrates some of the limitations of standard EIA practices and the potential of geomorphological inputs for improving those assessments. EIA is normally applied as a predictive tool, to assess (forecast) the effects of individual projects. However, some important impacts cannot be properly assessed at project scale; rather, they are the result of a set of actions which could be better considered within a land-use or urban expansion plan. These plans (or lack of them) determine land occupancy patterns and the impacts derived from them.

In the Rio Acre basin the indiscriminate defforestation and occupation on the alluvial belt produced a relatively small impact as a result of "active" human interference in river dynamics; this however, gives a false impression of the overall impacts. The large impact produced in the urban area by the slightly greater magnitude of the 1997 flood is the consequence of this process. The "passive" human intervention actually played a more significant role on the final impacts. Unplanned urban expansion, with a marked increase in the settlements exposed to flooding, was the main cause of the much greater damages experienced in 1997.

We must conclude that policies of urban and environmental planning ought to be developed with these ideas in mind. A wider concept of EIA should be applied, and "soft" measures for impact mitigation, such as disaster prevention and environmental education programmes, implemented.

But better planning of "passive" elements with respect to river dynamics is not sufficient. Up to now, annual deforestation rate in the state of Acre has been relatively low. However, deforested areas are concentrated along the Rio Acre basin, particularly between the river and the south-western water divide (Fig. 8). If this trend continues, impacts such as increased runoff and erosion and consequent increase in sediment load and channel metamorphosis ought to be expected; higher flood frequency and/or intensity would (can) be one of the results. Thus, deforestation should not only be considered in terms of its impacts on biota and biodiversity, but of geomorphological processes as well. The service which forests provide as regulators of fluvial dynamics is of paramount importance in tropical areas. The type of "active" interference described can seriously hamper the ability of the environment to provide such service and have important consequences for human well-being, as illustrated above.

The situation described shows that sustainable resource exploitation and urban development cannot be achieved without an appropriate assessment of geomorphological impacts, which can be best performed through plans, rather than projects.

AKNOWLEDGEMENTS

We thank ANEEL (National Agency of Electric Power) for stage and discharge data from Rio Branco and Brasileia gauging stations.

REFERENCES

FUNTAC (Technological Foundation of the Acre State). 1989.Deforestation in the Acre basin. *FUNTAC, (unpublished report)*, 34p, Rio Branco.

Gibs, R.J. 1967. The Geochemistry of the Amazon River system: part I. The factors that control the salinity and the composition and concentration of suspended solids. *Geological Society of American Bulletin*, 78: 1203-1232.

IBGE (Instituto Brasileiro de Geografia e Estatística). 1991. *Censo Demográfico, I*. Rio de Janeiro.

Leopold, L.B., Wolman, M.G., & Miller, J.P. 1964. *Fluvial processes in Geomorphology*. San Francisco, W.H. Freeman Co.

Mertes, L. 1985. Floodplain development and sediment transport in the Solimões-Amazon river, Brazil. Master Thesis, University of Washington.

Nordin, C.F. and Meade, R.H. 1982. Deforestation and Increased flooding of the upper Amazon. *Science* 215: 246-247.

Prefeitura Municipal de Rio Branco. 1986. Primeiro Plano Diretor de Desenvolvimento Urbano. Rio Branco.

Salati, E., Marques, J. & Molion, L.C. 1978. Origem e distribuiçao das chuvas na Amazonia. *Interciencia* 3: 200-205.

Walker, I. 1991. Algumas Consideraçoes sobre um programa de Zoneamento da Amazonia. Bases Científicas para Estratégias de preservaçao e Desenvolvimento da Amazonia: fatos e perspectivas. In: Val, A. Figliuolo, R. & Feldberg, E. (eds.). INPA, Manaus: 37-46.

Geomorphological impacts and feasibility of laying-out ski-trails in a glacial cirque in the Vosges (France) using GIS and multi-criteria analyses

S.J.E. van Dijck
University of Utrecht, The Netherlands

ABSTRACT: A method is proposed for the assessment and evaluation of impacts on landforms and soil erosion on ski-trails in an ancient glacial cirque in the Vosges (France). The impacts are quantified using a Geographical Information System (GIS) technique and an empirical soil-erosion model. Four alternatives for ski-trails are investigated with particular respect to geomorphology and soil loss. The most favourable alternative location proposed for ski-trails is determined by comparing the impacts with aspects of feasibility using a decision-support system.

1 INTRODUCTION

A plan was put forward to lay out ski-trails in the ancient cirque of Faignes-Fories in the mid-Vosges (France) (CERREP 1994a, b). Four alternative locations were considered for the skitrails (Fig. 1). A preliminary comparison of these alternatives by the property developer, considering cost-benefit ratio and impacts on vegetation, fauna and tourist appeal, identified the second alternative as the best (CERREP 1994a, b). This study assesses some geomorphological impacts of the laying-out of the ski-trails and compares these with some aspects of the feasibility of the project that depend on relief and substrate. Geomorphological impacts are defined here as those changes in landscape processes and landforms brought about by a project that are 'sensed' by one or more interest groups in society.

Two main modifications of the soil surface are required for the laying-out of the ski-trails. The first modification consists in the forest clearance. The substitution of trees by a grass cover will cause a decrease in interception storage. Soils on the ski-trails will be less permeable than forest soils, due to the absence of a litter layer and compaction by skiers. Both effects may result in a greater rainfall excess available for overland flow and soil erosion. The second modification consists in earth movings becouse soil and rock will have to be removed from places and to be deposited at others for the denivellation of the soil surface. This will affect the landforms of the cirque. Soil loss and the affection of those landforms are the geomorphological impacts under consideration in this study. In addition, the feasibility of the project is evaluated according to the accessibility of the area for machinery and the labour-intensiveness of preparing the different types of substrate for the laying-out of the skitrails.

The areas affected by and the volumes of soil concerned with the geomorhogical impacts and feasability of the project were calculated for each of the four alternative locations of the ski-trails using a GIS (ILWIS, version 1.4) (ITC 1994) and the soil-erosion model of Morgan et al. (1984). A digital elevation model and maps of the substrate, land use or vegetation, roads and trails, and constituent landforms of the cirque were used in the GIS. From these, new maps were calculated with tables in which map layers were combined by specifying keys. The spatial resolution of these maps was 25 m. The alternative with the lowest impact and highest feasibility was determined with the decision-support system DEFINITE (Jansen & Herwijnen 1992).

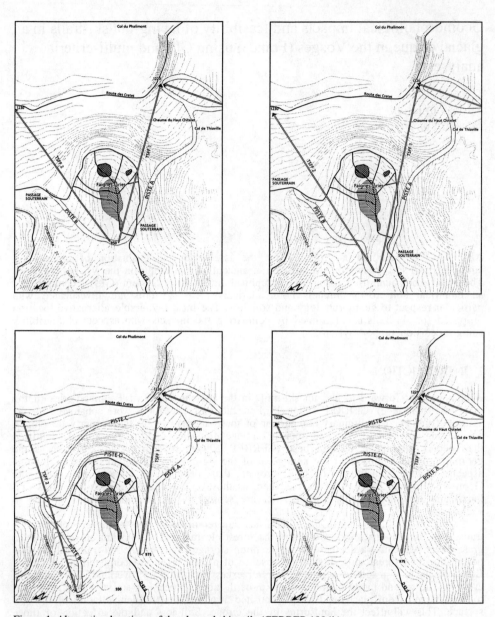

Figure 1. Alternative locations of the planned ski-trails (CERREP 1994b)

2 SUBSTRATE AND GEOMORPHOLOGY

The glacial cirque of Faignes-Fories is situated in the head of the valley of the River Vologne, at the west side of the Massif of the Hohneck in the Vosges (France). The morphological elements which constitute the cirque are the head wall, the bottom and the end moraine (Fig. 2). The end moraine, closing the cirque, overlays a treshold of bedrock. The end moraine is considered to mark one of the phases of readvance during the retreat of the glacier covering the cirque in the Late-Glacial.

The treshold is dissected by the stream of Faignes-Fories. The cirque is carved into a porphyritic granite, which is exposed at the steep southern part of the cirque. Traces of the loosening of blocks by freezing and thawing at the rim of the cirque and of the polishment of the bedrock are found on the head wall. At the foot of the wall, colluvial deposits are found superposed on morainic deposits in the bottom of the cirque (Fig. 3). The central part of the cirque bottom is occupied by a peat bog. The morainic deposits in the cirque consist of blocks up to three metres in cross section in a matrix of sandy loam or loamy sand.

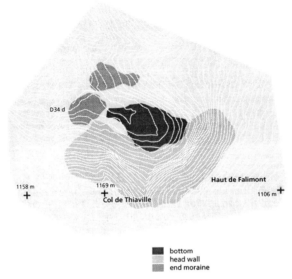

Figure 2. Morphological elements of the cirque of Faignes-Fories.

3 GEOMORPHOLOGICAL IMPACTS

3.1 *Soil loss*

Several studies of soil erosion and surface runoff on ski-trails in Switzerland and France have revealed that these processes are enhanced on ski-trails in comparison with the undisturbed soil (Bayfield 1974, Mosimann 1981, 1983, 1985, Giessübel 1988). Among the responsable modifications of soil and soil surface are cited: an increased bulk density, a reduced vegetation cover and a reduced rooting depth.

The impact on soil erosion in the cirque of Faignes-Fories due to the laying-out of the ski-trails was evaluated using the soil-erosion model of Morgan et al. (1984). The model does not describe the transfer of overland flow and soil in space. It was chosen because it requires only a small number of input parameters and because it is easily used in combination with a raster GIS, in contrast to physically-based models. The model was used to predict annual soil loss in 25 m² raster cells for the actual state of the soil surface (pre-project stage), and for the state with the ski-trails (post-project stage). There are no data of the annual soil loss in the cirque for the judgement of model permorfance. The model could therefore only be used to compare the annual soil loss in the pre- and post-project stages.

The model of Morgan et al. (1984) seperates the soil erosion process into a water phase and a sediment phase. The sediment phase is based on the scheme described by Meyer & Wischmeier (1969). It considers soil erosion to result from the detachment of soil particles by raindrop impact and the transport of those particles by overland flow. Thus the sediment phase comprises two equations: one for the rate of splash detachment and one for the transport capacity of overland flow (equations 1, 2). Sources of the values for the model parameters are given between parentheses. LDB refers to a local database, M to measured values, GIS to values provided by the GIS and REF to guides values taken from literature.

$$F = K (E\, e^{-aA})^b\, 10^{-3} \tag{1}$$

$$G = C Q^d \sin S \; 10^{-3} \tag{2}$$

where F = rate of soil detachment by raindrop impact (kg/m²), K = soil detachability index (g/J) defined as the weight of soil detached from the soil mass per unit of rainfall energy (M), E = kinetic energy of rainfall (J/m²), A = percentage of rainfall contributing to permanent interception and stemflow (LDB), $a = 0.05$ (Morgan 1996), $b = 1.0$ (Morgan 1996), $d = 2.0$ (Morgan 1996), G = transport capacity of overland flow (kg/m²) crop cover management factor (combines C and P factors of the Universal Soil Loss Equation to give ratio of soil loss under a given management to that from bare ground with downslope tillage, other conditions being equal (Expression of surface roughness, REF)), Q = volume of overland flow (mm), S = steepness of the ground slope expressed as the slope angle (GIS).

The inputs for equations 1 (rainfall energy equation) and 2 (runoff volume equation) are obtained from the water phase (equations 3-6):

$$E = R (11.9 + 8.7 \log I) \tag{3}$$
$$Q = R \exp(-R_c/R_0) \tag{4}$$
$$R_c = 1000 \; MS \; BD \; RD \; (E_t/E_0)^{0.5} \tag{5}$$
$$R_0 = R/R_n \tag{6}$$

where: R = annual rainfall (mm) (LDB), I = typical value for intensity of erosive rain (mm/h) (11 for temperate climates according to Morgan (1996)), MS = soil moisture content at field capacity or 1/3 bar tension (vol.%) (LDB), BD = bulk density of the top soil layer (kg/m³) (LDB), RD = top soil rooting depth (m) defined as the depth of soil from the surface to an impermeable or stony layer; to the base of the A horizon, to the dominant root base; or to 1.0 m, whichever is the shallowest (M), E_t/E_0 = ratio of actual (E_t) to potential (E_0) evapotranspiration (LDB), R_n = number of rain days in the year (LDB).

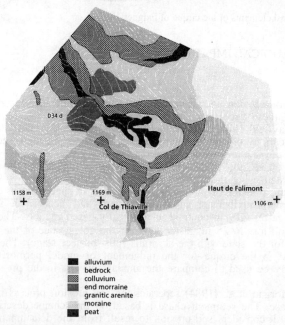

Figure 3. Morpholithological map of the cirque of Faignes-Fories.

The model compares the predictions of detachment by rainsplash and the transport capacity of the runoff and assigns the lower of the two values as the annual rate of soil loss, thereby denoting whether detachment or transport is the limiting factor.

Values for the parameters for the Morgan et al. (1984) model are given in Table 1. Values for BD and RD used to represent the ski-trails on the alternative locations are based on observations and samples from an existing ski-trail in the cirque. Values for soil detachability (K) were calculated from the encarvement of forest trails and their period of existence (approximately 30 years).

Table 1. Values of parameters in the Morgan et al. (1984) method of predicting soil loss in the cirque of Faignes-Fories.

Substrate	Land use	BD* (kg/m³) mean	st. err.	MS* (vol%) mean	st. err.	RD (m)	K (g/J) x10^4	A** (%)	E_t/E_0**	C***	R**** (mm)	R_n****
Granite	forest	0.79	0.13	0.37	0.06	0.4	2.1	30	0.9	0.001	1858	203
Granitic arenite	meadow	0.16	0.12	0.10	0.03	0.1	0.5	10	0.9	0.010		
Moraine	forest	1.03	0.13	0.21	0.05	0.4	2.8	30	0.9	0.001		
Colluvium	forest	1.08	0.14	0.22	0.06	0.4	2.8	30	0.9	0.001		
Peat	forest	0.82	0.18	0.42	0.06	0.4	2.1	30	1.0	0.001		
Granitic arenite	ski-trail	1.52		0.10	0.02	0.1	0.5	5	0.9	0.005		
Granitic colluvium	ski-trail						2.8	5	0.9	0.005		
Moraine colluvium	forest track	1.52		0.10	0.02	0.0	2.8	0	0.1	1.000		

* Reutenauer (1987)
** Biron (1994)
*** Morgan et al. (1984)
**** Météo-France (column is not filled out)

Input values for the pixels on the planned locations of the ski-trails, representing the situation after the laying-out of the ski-trails, reflect an increased bulk density (BD), a reduced rooting depth (RD), a reduced interception storage by vegetation (A) and a decreased surface roughness (higher values of C).

Model calculations were performed for the cirque in its actual state (pre-project stage) and for the four states with the proposed locations for the ski-trails (post-project stage). The total amounts of annual soil loss on the locations of the ski-trails in the post-project stage were standardized for the corresponding values in the pre-project stage.

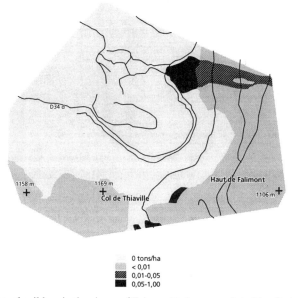

Figure 4. Actual annual soil loss in the cirque of Faignes-Fories as predicted by the Morgan et al. (1984) model.

As an example of the output of the Morgan et al. (1984) model, the map of the actual annual soil loss is shown in Figure 4. Values of annual soil loss are nowhere higher than 1.00 tons/ha. Assuming a bulk density of 1.03 kg/m^3 this corresponds to 0.01 mm of soil removed in a year. This negligible loss is in line with the belief that soil loss is not significant in the Vosges (Mercier, pers. comm.), at least not outside the existing ski-trails and besides there are no data on rates of soil loss on ski-trails in the Vosges.

The error in the predicted values for the actual annual soil loss was calculated by adding and subtracting the amount of one standard error to or from each of the input values for the parameters. The standard errors were calculated from the parameter values or estimated. The standard errors were either added or substracted in the GIS in order to obtain a lower and upper limit of the actual value of annual soil loss. The map of the actual annual soil loss was subtracted from the map of the upper limit. The errors are in between 10 and 20% of the predicted values for the actual annual soil loss.

3.2 *Impacts on the morphology of the cirque of Faignes-Fories*

In this study, impacts on the morphological elements of the cirque (head wall, bottom, end moraine) are defined as the percentage of the areal extent of the elements covered by ski-trails, and also as the percentage of volume extracted from the end moraine. The four alternatives were compared with respect to these values. In the calculation of the areal extent the true area of each pixel under consideration was taken into account by correcting for the slope of the pixel.

The volume of soil to be removed from the end moraine for denivellation of the soil surface was estimated as follows. The values of the digital elevation model were lowered by 0.5 m, which could be the mean soil thickness to be affected by denivellation. The lowered DEM was smoothed by filtering the map using a standard filter in ILWIS. This filter assigns the central pixel of a moving kernel the mean value of all pixels in the kernel. Then for each pixel the height difference between the DEM and the lowered and smoothed DEM was calculated. This height difference represents the estimated soil thickness to be removed for denivellation. For each pixel the height difference was multiplied by the true area of each pixel to obtain the volume of soil to be removed from each pixel. These values were summed for all pixels occurring on each alternative ski-trail.

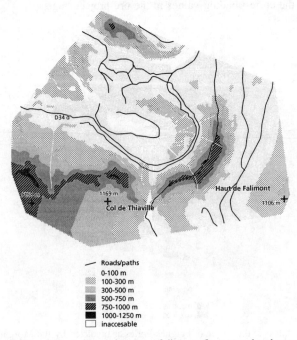

Figure 5. Accessibility of the study area in terms of distance from paved and unpaved roads.

4 SOME ASPECTS OF FEASIBILITY

The feasibility of a project is defined as the extent to which it can be implemented. The most important aspects of feasibility for this project with respect to relief and substrate are the accessibility of the project area for machinery and the labour-intensiveness of the preparation of the substrate.

The accessibility of the study area for machinery is expressed as the distance (m) from the paved and unpaved roads (Fig. 5), which are passable by a bulldozer. For the calculation of the distance in ILWIS roads were defined as 'source areas'; streams and humid sites were defined as 'inaccessible'. ILWIS calculates distances from sources to inaccessible sites by interpolation. Areas with slopes steeper than 40% in between sources and inaccessible sites were assigned a friction factor 10.

For the denivellation of the soil surface trees have to be chopped and large boulders removed. If the substrate is blocky, fine-grained material has to be applied to create a soil with smaller pore sizes and to smooth the surface. In general the humus layer at the surface is removed during denivellation. It has to be reapplicated to increase the water retention capacity and cohesion of the soil (Mosimann 1981,1983,1985, Giessübel 1988). If no humus layer is present at the surface the humus has to be derived from elsewhere.

The types of substrate in the study area were ranked with respect to the labour-intensiveness of carrying out the above-mentioned modifications (Tab. 2). The ranking scores of all pixels occurring on each alternative location of the ski-trails were summed using the GIS.

Table 2. Degree of labour-intensiveness (1 is lowest, 6 is highest degree) of removing or applicating blocks, soil or humus for denivellation.

Substrate	Degree
granitic arenite	1
moraine	2
peat	3
colluvium	4
end moraine	5
granite bedrock	6

To granite bedrock was assigned the highest degree of labour-intensiveness: after the large blocks have been removed, the bedrock has to be blown up to create an artificial regolith. As the bedrock is found on the steepest slopes in the cirque (up to 55°), the products of the degradation processes may not remain on the slope as desired. Humus has to be derived from elsewhere to cover the artificial regolith, because no or only a thin humus layer is found on this substrate (0-5 cm).

The soils in the peat bog (Fig. 3) have a low soil strenght because, due to their low position in the cirque, they are permanently saturated. For the laying-out of ski-trails on these soils material would have to be applied to increase the bearing capacity of the soils. They will however continue to accumulate water because of their terrain position, unless the flow routing of the supplying streams is changed.

The colluvium in the study area consists of medium sized blocks (25 cm in cross section on average). The content of sands and fines is low in comparison with the granitic arenites and the moraines. On the hillslopes the colluvium acts as a sieve for water supplied from upslope, and therefore constitutes the most unstable type of unconsolidated substrate in the study area. Humus layers are absent or very thin. For denivellation sands and fines have to be added to create a soil capable of storing water and to smooth the soil surface. Humus has to be supplied from elsewhere.

The end moraine is characterized by the largest block size (up to 3 m in cross section) compared to the other types of moraine. It is piled up on the bedrock treshold, which has to be blown up after the removal of the blocks. To the end moraine was therefore assigned a higher degree of labour-intensiveness. The soils in the granitic arenites on the hilltops and fringing faint slopes (Fig. 3) constitute the best substrate for the laying-out of the ski-trails. No large blocks have to be removed, no regolith has to be created and no humus to be supplied, since the humus layer in the soils is usually about 20 cm thick.

5 COMPARISON OF THE ALTERNATIVE LOCATIONS FOR THE SKI-TRAILS USING MULTI-CRITERIA ANALYSES

The geomorphological impacts and feasibility of the plan to lay-out ski-trails in the cirque of Faignes-Fories were evaluated using the decision support system DEFINITE (Janssen & Herwijnen 1992). DEFINITE supports decision making on cases with finite sets of alternatives. The alternatives are the four locations of the ski-trails, referred to below as variants 1 to 4.

5.1 Problem definition

The problem to be analyzed in DEFINITE was defined in an 'effects table'. An effects table includes the names of the alternatives (the column headings) and the names of the effects (the row headings). The effects are the geomorphological impacts and aspects of feasibility described above.
The values in Table 3 represent the magnitude of the effects for each alternative. Negative scores denote costs, positive scores benefits. Thus a higher score on accessibility is considered as a benefit, implying that an alternative location is more easily accessible from and nearer to roads and paths. The effect scores on the geomorphological impacts and the aspects of feasibility of the the four alternatives are given in Table 3.

Table 3. Effects table of geomorphological impacts and aspects of teasibility.

Alternative Effect	Variant 1	Variant 2	Variant 3	Variant 4
Ratio of total annual soil loss on ski-trails post-/pre-project stage	-16.1	-20.8	-7.5	-6.3
Surface affection morph. elements. (m^2)	-23,000	-25,000	-52,000	-52,000
Extraction from end moraine (m^3)	-5699	-6793	-1131	-1131
Preparation of substrate (summed scores)	-8424	-9917	-14,000	-13,000
Accessibility (< 100 m/m^2)	52,000	69,000	77,000	64,500

5.2 Problem evaluation

The effect scores were evaluated using several multicriteria methods. The alternatives were compared and ranked according to their scores on all effects under consideration. The multicriteria methods used for the evaluation of impacts and feasibility are the Weighted summation method, the Electre II method, the Regime method and the Evamix method. All methods generate a ranking of the alternatives. The use of the methods requires information on the relative importance of each effect: weights need to be assigned. In the Regime method it is sufficient to order the effects from the most to the least important. All other methods require quantitative weights. In DEFINITE, weights can be assigned directly or by using methods that transform qualitative priority statements into quantitative weights (Beinat 1994). Here weights were assessed using the expected value method. In this method, weights are assigned to the effects according to their importance in a preliminary qualititative priority statement. All weights sum to 1. Two weights sets were created from the viewpoint of respectively a conservancy foundation and a property developer: one emphasizing the importance of the geomorphological impacts, the other emphasizing the importance of the aspects of feasibility of the project (Tab. 4).

Table 4. Weights of geomorphological impacts ((GI) and aspects of feasibility (F) assessed using the expected value method.

Ranking		Weight	
		GI	F
1	annual soil loss ratio (post/pre-project)	0.547	0.157
2	surface affection of morphological elements	0.257	0.040
3	volume extracted from end moraine	0.157	0.090
4	accessibility	0.090	0.457
6	surface preparation	0.040	0.257

Accelerated soil loss in the cirque on the ski-trails was considered as the most important geomorphological impact because it would imply the degradation of soils and vegetation on the trails and an increased supply of sediment to the peat bog in the bottom of the cirque.

As numerous cirques are found in the Vosges (Tricart & Cailleux, 1967), the surface affection of the morphological elements might be considered of minor importance. The extraction of material from the end moraine was given the lowest weight of the geomorphological impacts, because this effect concerns only one of the morphological elements of the cirque.

The accessibility of the locations of the ski-trails from roads and trails was considered a more important effect than the labour-intensiveness of the surface preparation, simply because the latter would not be practicable without machines having access to the sites. Accelerated soil loss on the ski-trails will undo their implementation in the long term. It was therefore considered more important for the feasibility of the project than the volume to be extracted from the end moraine and the surface affection of the morphological elements.

Before a multicriteria method can be applied, the effects table needs to be standardized. In the present case, interval standardization was used. The scores were standardized for each effect with the following formula:

$$S_e = (S_{ij} - S_{min}) / (S_{max} - S_{min}) \tag{7}$$

where S_e = standardized effect score, S_{ij} = effect score, S_{min} = lowest score, S_{max} = highest score.

The standardized scores received values between 0 and 1. They were used for calculating the ranking of the alternative locations of the ski-trails with the four multicriteria methods.

Weighted summation is a simple and frequently used multicriteria method. An appraisal score is calculated for each alternative by first multiplying each standardized effect score by its appropriate weight, followed by summing the weighted scores of all effects:

$$A_j = \sum_{i=1}^{k} w\, S_e \tag{8}$$

where A_j = appraisal score for alternative j, S_e = standardized effect score, w = weight, k = number of effects (i=1, 2, ...,k).

The Electre II method calculates the ranking of the alternatives based on pairwise comparison of the alternatives. No ranking scores are obtained. Both the degree to which one alternative is better than the other alternatives (concordance) and the degree to which it is worse than other alternatives (discordance) are used. 'Strong' and 'weak' treshold values can be entered, which indicate to what extent an alternative should at least and at most be better or worse than another to achieve a ranking. Values for the concordance-strenght should be between 0.5 and 1 (1 is the strongest value). Values for the discordance should be between 0 and 1 (0 is the strongest value). In the present case, treshold values were assigned automatically by the algorithms provided by the decision support system (Tab. 5).

Table 5. Treshold values for the ranking of alternatives with the Electre II method. C(ij) and D(ij) are the weights corresponding to the effect i for which alternative j is better respectively worse than another alternative, n_i is the number of effects and n_j the number of alternatives.

Concordance-index	strong	$\Sigma\Sigma\, 1.1\, C_{(i,j)} / (n_i-1)(n_j-1)$
Concordance-index	weak	$\Sigma\Sigma\, 0.9\, C_{(i,j)} / (n_i-1)(n_j-1)$
Discordance-index	strong	$\Sigma\Sigma\, 1.1\, D_{(i,j)} / (n_i-1)(n_j-1)$
Discordance-index	weak	$\Sigma\Sigma\, 0.9\, D_{(i,j)} / (n_i-1)(n_j-1)$

The Regime method uses a random generator for each pair of alternatives to estimate the probability that one alternative ranks above the other. Qualitative weights were entered correspondingly to the ranking of the effects (Table 4).

The Evamix method generates separate dominance tables for quantitative and qualitative effect scores. Both tables are combined into a total dominance table. As none of the effects among the geomorphological impacts and aspects of feasibility was measured on a qualitative scale, only quantitative dominance tables were created.

The rankings of the alternatives using all methods are shown in Figure 6 for the case in which more importance is attached to the detriment of the geomolphological impacts due to the laying-out of the ski-trails than to either the benefit of a high degree of accessibility or the labour-intensiveness of the preparation of the substrate. They suggest that the second location proposed for the ski-trails (variant 2) would be the most favourable alternative and the first location (variant 1) the second best. If in contrast the feasibility of the laying-out of the ski-trails is considered more important, variant 2 would be the second best alternative together with variant 4 (Fig. 7). Variant 2 was promoted by the property-developer in a preliminary comparison of the alternatives, considering cost-benefit ratio and the impacts on vegetation, fauna and tourist appeal. Nonetheless variant 1 would be the best alternative according to all but the Electre II methods. If only this method had been used, the best alternative would be variant 2 for as well the least detriment of geomorphological impacts and, together with variant 4, for the benefit of feasibility. Obviously different multicriteria methods can generate different rankings of alternatives, using the same set of weights on the effects. It is therefore advisable to use more than one method.

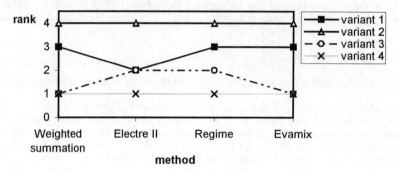

Figure 6. Ranking of alternatives with emphasis on geomorphological impacts.

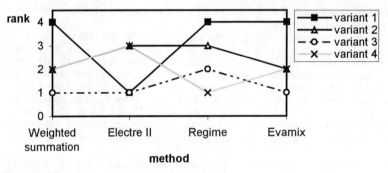

Figure 7. Ranking of alternatives with emphasis on feasibility.

5.3 *Sensitivity analysis*

The rankings which result from the multicriteria analyses are subject to uncertainty, associated with the type of method, the effect scores and the weights used. Before a decision is taken in favour of one of the alternatives on the basis of the rankings, this uncertainty should be analyzed. The uncertainty in the rankings due to the use of different multicriteria methods was discussed before (see 5.2). The uncertainty resulting from the effect scores and weights was analyzed with DEFINITE.

The uncertainty in the effect scores is expressed as the percentage that the actual values of the effect can be maximally higher or lower than the scores included in the effects table. We assumed the uncertainty in the scores on the post- to pre-project ratio of annual soil loss predicted by the Morgan et al. (1984) soil-erosion model to equal the maximum error in the predictions: 20% (see 3.1). The uncertainty in the effects expressing areas of morphological elements of the cirque covered by future ski-trails (surface affection, extraction of end moraine) is the result of the ski-trails becoming wider or smaller when rasterized. The ski-trails are in between 20 and 40 m width, and may be covered by one or two pixels of 25 m width. The maximum absolute error in the width of the affected areas on the rasterized and crossed maps of the morphological elements and the ski-trails would then be the difference between the maximum possible width covered (2 x 25 m) by pixels and the minimum true width of the areas (20 m), that is 30 m. Expressed as a percentage of the average width of the areas affected this represents an error in the effect scores of 100%.

The uncertainty in the areas at a distance of 100 m or less to roads and forest trails was estimated as one pixel length from the limit of the 100 m zones bordering the roads and trails on each side. Accordingly the uncertainty in the effect scores on accessibility is 50/200 m/m or 25%. In contrast with the scores on the other effects, the effect scores on the labour intensiveness of the preparation of the soil surface were measured on an ordinal scale. They are the sums of ordinal values attached to each type of substrate occurring on the future ski-trails (Tab. 2). Where a ski-trail crosses two adjoining units on the lithogenetic map, the pixels along the border between the units were considered to be possibly assigned the degree of labour-intensiveness corresponding with the other lithogenetic unit. An average length of two pixels was assumed for the crossing between the ski-trail and the border between the map units. The uncertainty in the effect score on labour-intensiveness for the ski-trail was then calculated as the summed differences between the degrees of adjoining lithogenetic map units for average border lengths of two pixels, as a percentage of the summed degrees. The percentages for the four variants are between 0.46 and 0.50. The final rankings of the alternatives appeared to be insensitive to the uncertainty in the effect scores, whichever the multicriteria method used.

The uncertainty in the weights is expressed as the percentage that the actual weight value of an effect can be maximally higher or lower than the weight included in the weights set (Tab. 4). We already accounted for an uncertainty in the weights by creating two sets of weights with higher values on either the geomorphological impacts or the aspects of feasibility, in correspondance with the viewpoints of respectively a conservancy foundation and a property developer. We assumed values for the uncertainty of all weigths of successively 5, 10, 20 and 100%. If the weights set with the emphasis on the geomorphological impacts was used, no changes occurred in the rankings of the alternatives. For the weights set with the emphasis on the aspects of feasibility, the only change was the ranking of variant 2 above instead of equal to variant 4. This change was obtained for an uncertainty in the weights of 5, 10 and 20% when the Evamix method was used, an for an uncertainty of 100% in the weights when the Weighted summation method was used. The rankings of the first and second best alternatives were not influenced by any uncertainty in the weights.

6 SUMMARY AND CONCLUSIONS

Laying-out ski-trails in the cirque of Faignes-Fories will change actual landscape processes and landforms. If these changes are 'sensed' by an interest group in society, like a conservancy foundation, they are termed 'geomorphological impacts'. Some geomorphological impacts of this project are soil loss on the future ski-trails and the covering with the ski-trails of parts of the bottom, head wall and end moraine, which are the constituent elements of a glacial cirque. Four alternative locations were proposed for the ski-trails. A conservancy foundation would opt for the one with the least geomorphological impact.

In order to lay-out the ski-trails soil and rock need to be moved with the help of machinery, the ease of which depends on the accessibility of the project site from roads and forest trails, and on the labour-intensiveness of preparing the substrate. These are the most important aspects of the feasibility of the project, constrained by the geomorphology of the cirque in terms of relief and substrate. In contrast with the conservancy foundation, the property developer would find benefit in the most feasible alternative location of the ski-trails.

The geomorphological impacts and aspects of feasibility were quantified with the help of a raster-GIS. The input maps required were a digital elevation model, maps of the substrate, the land use or vegetation and of roads and forest trails, and maps with the alternative locations of the ski-trails. Some of the GIS functions used are for instance the slope and distance functions for calculating respectively the steepness of the ground surface as one of the inputs of the soil-erosion model, and the distance of each pixel in the cirque from roads and forest trails for quantifying the accessibility of the project sites.

The quantified geomorphological impacts and aspects of feasibility could be used, with the help of a decision-support system, to decide one of the alternative locations for the ski-trails. The system enables the execution of one or more multicriteria analyses. In this study the criteria (or effects) were the geomorphological impacts and the aspects of feasibility. The relative importance of the effects was expressed by weights. Two sets of weights were used for expressing the relative importance of the effects from the viewpoints of a conservancy foundation and the property developer. The ranking of alternatives resulting from the analyses appeared to be sensitive to the methods chosen and to the weigths imposed on the effects, but not to uncertainties in the values of the geomorphological impacts and aspects of feasibility.

Three multicriteria methods ranked the second alternative location of the ski-trails as the best from the viewpoint of the conservancy location, and the first alternative location from the viewpoint of the property developer. Obviously users of the decision-support system with contrasting interests would obtain different rankings of the alternatives because they attach importance to different effects, and would consequently decide to different alternatives. An unbiased decision could only be taken if all weights were put to equal values. If only the Electre II method had been used, the best alternative would be the second alternative from the viewpoint of the conservancy foundation and, together with variant 4, also for the benefit of the property developer. The use of only one multicriteria method can obviously result in a ranking of alternatives that would not be obtained when using more methods. More methods should be used to verify the obtained rankings of alternatives.

ACKNOWLEDGEMENTS

I kindly thank Daniel Blumenroeder and Isabelle Eglin of the Faculty of Geography of the University of Louis Pasteur in Strasbourg (France), Daniel Blumenroeder for his never ceasing technical support and enthousiasm and Isabelle Eglin for her advise, cordiality and help in the laboratory. I especially thank Ton Markus who was able to make my maps comprehensible and did this in his free time.

REFERENCES

Beinat, E. 1994. *Methods for Environmental Impact Assessment. Documentation of course on Environmental Impact Assessment.* Amsterdam: Tinbergen Institute, Institute for Environmental Studies, Free University 28.

Bayfield, N.G. 1974. Burial of vegetation by erosion debris near ski-lifts on Cairngorm, Scotland. *Biological Conservation* 6: 246-251.

Biron, Ph. 1994. Le cycle d'eau en forêt de moyenne montagne: flux de sève et bilans hydriques stationnels (bassin versant du Strengbach à Aubure - Hautes Vosges). Thèse de doctorat de l'Université Louis Pasteur. Strasbourg: Centre d'Etudes et de Recherches Eco-Géographiques.

CERREP S.A. 1994a. *Unité Touristique Nouvelle: note de présentation.* La Bresse (Dpt.Vosges): Meylan.

CERREP S.A. 1994b. *Comparaison des variantes de la Liaison entre les domaines skiables.* La Bresse (Dpt.Vosges): Meylan.

Giessübel, J. 1988. Nützungsschäden, Bodendichte und rezente Geomorphodynamik auf Skipisten der Alpen und Skandinaviens. *Zeitschrift fur Geomorphologie* 70: 205-219.

ITC 1994. *User manual of the Integrated Land and Water Information System version 1.4.* Enschede: International Institute for Aerospace Survey and Earth Sciences.

Janssen, R. & Van Herwijnen, M. 1992. *DEFINITE, Decisions on a finite set of alternatives - Instructions for use of demonstration disks.* Dordrecht: Kluwer Academic Publ.

Meyer, L.D. & Wischmeier, W.H. 1969. Mathematical simulation of the process of soil erosion by water. *Trans. of the Am. Soc. of Agr. Eng.* 12: 754-762.

Morgan, R.P.C., Morgan, D.D.V. & Finney, H.J. 1984. A predicive model for the assessment of soil erosion risk. *Journ. of Agr. Eng. Res.* 30: 245-253.

Morgan, R.P.C. 1996. *Soil erosion and conservation.* Essex: Longman.

Mosimann, Th. 1981. Geoökologische Standortsindikatoren für die Erosionsanfälligkeit alpiner Hänge nach Geländeeingriffen für Pistenanlagen. *Geomethodica* 6: 143-174.

Mosimann, Th. 1983. Landschaftsökologischen Einfluss von Anlagen für den Massenskisport, II Bödenzustand und Bödenstörungen auf planierten Skipisten in verschiedenen Lagen (Beispiel Crap Sogn Gion, Laax). In H. Leser (ed), *Materialen zur*

Physiogeographie, Basler Beiträge zur Physiogeographie. Basel: Geographisches Institut der Universität Basel, Ordinariat für Physiogeographie 3.

Mosimann, Th. 1985. Landschaftsökologischen Einfluss von Anlagen für den Massenskisport, III Ökologischen Entwicklung von Pistenflächen. Entwicklungstendenzen im Erosionsgeschehen und beim Wiederbewuchs planierter Pisten im Skigebiet Crap Sogn Gion / Laax GR. In H. Leser (ed), *Materialen zur Physiogeographie, Basler Beiträge zur Physiogeographie*. Basel: Geographisches Institut der Universität Basel, Ordinariat für Physiogeographie 9.

Reutenauer, D. 1987. Etude de la variabilité spatiale des propriétés physiques et hydriques des sols et des formations superficielles du bassin de la Fecht, en amont de Türckheim (Haut-Rhin). Thèse de doctorat de l'Université Louis Pasteur. Strasbourg: Centre d'Etudes et de Recherches Eco-Géographiques..

Tricart, J. & Cailleux, A. 1967. *Le modelé des régions périglaciaires, Tome II*. Paris: Sédes.

A methodology for assessing landscape quality for environmental impact assessment and land use planning; Application to a Mediterranean environment

L. Recatalá & J. Sánchez
Centro de Investigaciones sobre Desertificación-CIDE, Consejo Superior de Investigaciones Científicas, Universitat de València, Generalitat Valenciana, Albal (Valencia), Spain

ABSTRACT: An approach for organising and structuring the integration of geomorphological and other features for landscape description and assessment is presented.

The method described, based on the identification and mapping of integrated "mapping units", has been conceived for and applied in the coastal zone of the Mediterranean, but could easily be modified for adaptation to other environments. The method is based on the definition of four basic landscape properties: intrinsic and extrinsic landscape quality, intrinsic and extrinsic landscape fragility. These properties are combined to obtain landscape environmental quality.

The method proposed has been applied to an area in the province of Valencia. The results obtained show that categories of landscape value can be defined satisfactorily. These categories provide a useful basis for the evaluation of visual impacts and for the establishment of conservation priorities in the area.

1 INTRODUCTION

Fauna, flora, soil and landscape are important natural resources which can easily be modified by human activities. Landscape, like soil, can be considered as a non renewable resource which is not "consumed" by human use but can experience serious degradation (Gómez Orea 1978, Claver Farias et al. 1982, Villarino 1985, Westman 1985, Bolós et al. 1992). For this reason, it is important to include the consideration of landscape in the processes of land-use planning, environmental impact assessment (EIA) and conservation.

Landscape, can be considered, from the functional point of view, as a mosaic of "homogeneous units" characterised by their constitution and processes (Boluda et al. 1984, Ibáñez Martí et al. 1987), or from the point of view of perception or "visual image" (Zube 1974, Amir & Gidalizon 1990). The application of these two concepts results in the identification of different mapping units. In some cases, both types of concepts or considerations may apply (Bolós et al. 1992).

In this paper, we apply the latter concept of landscape. When it comes to homogeneous mapping units, in the functional sense, other terms, such as "ecosystem" or "environment" are probably more appropriate (González Bernáldez 1981). Hereafter, landscape refers to the visual image of a mapping unit. A mapping unit is then considered as a portion of land homogeneous for the components defining landscape (see below), on which an land-use decision is made.

Visual landscape quality is a vague and somewhat subjective concept. Moreover, the intrinsic quality of a landscape (from the visual point of view) is not the only characteristic to be taken into account for land-use planning and EIA purposes; the fragility of the landscape with respect to human interventions must also be considered. The overall "merit" or value (here named environmental quality) of a landscape should be the result of combining those two characteristics.

Three basic approaches have been proposed for the assessment of visual landscape value:

(1) Direct methods. Using these methods, landscape is visually assessed "in situ" by observers and /or by means of a photograph, a picture, etc. (Fines 1968, Carls 1974, Zube 1974, 1984, Garling 1976, Kaplan & Kaplan 1978).

(2) Indirect methods. In this case, landscape is assessed taking into account different landscape components or aesthetic categories which can be evaluated separately (Gómez Orea 1978, Amir & Sobol 1987, Amir & Gidalizon 1990).

(3) Mixed methods. These methods integrate both procedures. Thus, landscape is directly evaluated by observers taking into account several landscape components or aesthetic categories (Macia 1979).

All these methods have been criticized. Subjectivity is the major drawback of direct methods, because both the observer's experience (Cerny 1974, Carlson 1977, Escribano 1987) and the physical observation conditions (Litton & Tetlow 1974, Escribano 1987) influence the assessment. For indirect methods, the major problem comes from the selection of the most adequate and significant landscape components or landscape aesthetic categories (Dunn 1974, Arthur et al. 1977). Mixed methods have the problems encountered by both the direct and indirect methods and, furthermore, are difficult to develop and apply (Dunn 1974, Recatalá 1995). In fact, according to Dunn (1974) there is not an universally accepted method to assess visual landscape and the choice of a method depends on the objective pursued and the nature of the region.

The proposal presented here represents a modification of previous indirect methods, aimed at reducing the problem of subjectivity presented by direct methods and the complications of mixed methods. The difficulty of an adequate choice of components has been tackled, but not completely solved, through the use of sensitivity analyses. Antecedents of the method presented here can be found in Cendrero et al. (1986, 1990) and AMA (1989, 1991).

2 METHODOLOGY

Landscape environmental quality can be assessed through the separate consideration of two different properties: visual quality and visual fragility. Moreover, landscape environmental quality is influenced by the surrounding landscape within its visual basin (Ramos et al. 1976; Gómez Orea 1978; Claver Farias et al. 1982), a fact which is often overlooked (Recatalá 1995). In this proposal, the following visual landscape properties are considered for the assessment of environmental quality: intrinsic visual quality, intrinsic visual fragility, extrinsic visual quality and extrinsic visual fragility. The formers refer to the properties of a given unit and the latter to those of the surrounding units included in the visual basin. The visual basin of a given mapping unit includes all units totally or partially visible from the centre of that mapping unit. Techniques to define the visual basin of a mapping unit can be found, for instance, in Claver Farias et al. (1982).

The consideration of intrinsic and extrinsic quality and fragility is especially relevant for EIA studies. The effect of a certain action depends not only on the aesthetic value of a unit but also on its ability to integrate modifications and on the overall effect on the surrounding units. All of these, in turn, are to a great extent determined by geomorphologic characteristics.

2.1 *Intrinsic visual landscape quality*

This property refers to the value of a landscape to be preserved for its structure or significance (Gómez Orea 1978; Díaz de Terán 1985; Cendrero et al. 1986, 1990, González-Alonso 1995) and can be assessed taking into account several landscape components (relief, vegetation, etc.). Eight landscape components for assessing visual quality in Mediterranean regions are considered (Recatalá & Sánchez 1995b, Recatalá et al. unpubl.a): relief-gradient, relief-unevenness, altitude, lithology, presence of waterbodies, land use, vegetation and human constructions. The identification of these components was made on the basis of trial analyses in several pilot areas. A simple sensitivity analysis was carried out to evaluate the relevance of the identified components for explaining differences between the landscape visual quality of such areas (Recatalá 1995). Components were eliminated on a component by component basis from the expression used to evaluate intrinsic visual quality (see below) and their influence on the result analysed. Differences between the landscape visual quality of pilot areas were better explained when the eight components were combined. Several classes are established for each landscape component, considering the environmental characteristics of Mediterranean regions. They are the following:

1. Relief-gradient (RS).
    * RS1. Flat, gradient < 5%.
    * RS2. Gentle, gradient between 5% and 15%.
    * RS3. Hilly, gradient between 15% and 25%.
    * RS4. Prominent, gradient between 25% and 45%.
    * RS5. Steep, gradient > 45%.

2. Relief-unevenness (RU), which refers to differences in altitude within a mapping unit.
    * RU1. < 25 m.
    * RU1. Between 25 m and 75 m.
    * RU1. Between 75 m and 150 m.
    * RU1. Between 150 m and 300 m.
    * RU1. > 300 m.

3. Altitude (A), which refers to the absolute elevation of the unit above sea level.
    * A1. < 100 m.
    * A2. Between 100 m and 300 m.
    * A3. Between 300 m and 600 m.
    * A4. Between 600 m and 1000 m.
    * A5. > 1000 m.

4. Lithology (L).
    * L1. Alluvium and/or colluvium and/or marine sands.

    * L2. (a) Terrigenous carbonate rocks.
        (b) Clayey and gypsum rocks.
    * L3. (a) Terrigenous silicic rocks.
        (b) Clayey rocks.
    * L4. Limestones.
    * L5. (a) Sandstones and claystones.
        (b) Volcanic materials.

5. Waterbodies (W).
    * W1. No waterbodies.
    * W2. Intermittent streams or channels.
    * W3. Rivers or reservoirs.
    * W4. (a) Lakes.
        (b) Sea.
    * W5. Sea and lagoons.

6. Land use (LU).
* LU1. (a) Industrial-urban uses or infrastructures.
    (b) Dry stream or river beds.
* LU2. Annual crops.
* LU3. Fruit trees.
* LU4. Artificially re-forested areas.
* LU5. (a) Natural tree cover.
    (b) Wetlands.
    (c) Saltmarshes.
    (d) Beaches.

7. Vegetation (V).
* V1. (a) No vegetation.
    (b) Crops and/or other "artificial" vegetation.
* V2. Pasture.
* V3. (a) Scrub without trees.
    (b) Riparian vegetation without trees.
* V4. (a) Pasture or brushwood with trees.
    (b) Riparian vegetation with trees.
* V5. (a) Forest.
    (b) Vegetation of wetlands.
    (c) Vegetation of dunes.
    (d) Vegetation of saltmarshes.
    (e) Vegetation of cliffs.

8. Human Constructions (HC).
* HC1. Completely built-up areas (urban or industrial).
* HC2. (a) Partly built-up areas.
    (b) Quarries.
    (c) Dumps.
* HC3. (a) Rural zones with abundant constructions.
    (b) Rural zones with some high density developments.
* HC4. (a) Rural zones with scattered constructions.
    (b) Rural zones with some low density developments.
* HC5. Zones with no constructions.

The assessment of intrinsic visual quality through the above components was carried out using a weighting-rating approach. This requires that some conventional numerical values be assigned to the different classes or types of each landscape component. For instance, even numbers from 2 to 10 can be assigned to the different classes of each component; obviously, the greater the value, the higher the visual quality of the type, as shown in Table 1.

These values are, unavoidably, partly affected by the subjective of the authors. However, an independent test of their validity is provided by the fact that they coincide with other assessments carried out in the Valencia region (Diputación Provincial de Valencia, 1986; AMA, 1989, 1991).

Secondly, differential weights reflecting the importance of the different landscape components for visual quality must be established. A trial-and-error analysis was carried out using several combinations of weights (including the combination of all weights equal to 1) in several pilot areas. The most sensitive combination found (Recatalá 1995) was: relief-gradient = 2, relief-unevenness = 1, altitude = 1, lithology = 2, waterbodies = 2, land uses = 1, vegetation = 2 and human constructions = 1.

Table 1. Values assigned to the different types or classes of each landscape component for assessing intrinsic visual quality and intrinsic visual fragility.

Landscape components	Value for intrinsic visual quality	Value for intrinsic visual fragility
1. Relief-gradient		
RS1	2	10
RS2	4	8
RS3	6	6
RS4	8	4
RS5	10	2
2. Relief-unevenness		
RU1	2	2
RU2	4	4
RU3	6	6
RU4	8	8
RU5	10	10
3. Altitude		
A1	2	2
A2	4	4
A3	6	6
A4	8	8
A5	10	10
4. Lithology		
L1	2	2
L2a, b	4	4
L3a, b	6	6
L4	8	8
L5a, b	10	10
5. Waterbodies		
W1	2	2
W2	4	4
W3	6	6
W4a, b	8	8
W5	10	10
6. Land use		
LU1a, b	2	2
LU2	4	4
LU3	6	6
LU4	8	8
LU5a, b, c, d	10	10
7. Vegetation		
V1a, b	2	2
V2	4	4
V3a, b	6	6
V4a, b	8	8
V5a, b, c, d, e	10	10
8. Human Constructions		
HC1	2	2
HC2a, b, c	4	4
HC3a, b	6	6
HC4a, b	8	8
HC5	10	10
9. Accessibility		
AC1		10
AC2		8
AC3		6
AC4		4
AC5		2
10. P. Presence of observers		
O1		2
O2		4
O3		6
O4		8
O5		10

Intrinsic visual quality was then obtained using the following expression:

$$QVI_{pu} = \sum_{i=1}^{n} p_i x\, viq_{pu}$$

where QVI_{pu} = intrinsic visual quality of landscape p which is the visual image of mapping unit u, p_i = weight for landscape component i, n = number of landscape components, and viq_{ipu} = value of intrinsic visual quality of the type component i. The higher the QVI_{pu} value the greater the intrinsic visual quality of landscape p is.

QVI values have not a real meaning; they only serve for the purpose of comparing the relative intrinsic visual quality landscapes of different mapping units have and for establishing a rank, not a quantitative measure.

2.2 Intrinsic visual fragility

This property refers to the ability of the landscape to "integrate" or "conceal" human activities (Gómez Orea 1978, Díaz de Terán 1985, Cendrero et al. 1986, 1990, Escribano 1987). Intrinsic visual fragility may vary for different human activities, but it can be considered as a constant property for multi-purpose land-use planning; At project level, more detailed, activity-specific analyses are required (Díaz de Terán 1985, Escribano 1987). Thus, in landscape studies for EIA plans, intrinsic visual fragility -like visual quality- can be assessed in a generic way on the basis of several landscape components.

Ten landscape components for assessing intrinsic visual fragility in Mediterranean regions are proposed (Recatalá & Sánchez 1996b, Recatalá et al. unpubl.a). They are the same as for intrinsic visual quality plus accessibility and potential presence of observers. A simple sensitivity analysis (similar to the one made for intrinsic visual quality) was also carried out in this case to evaluate the relevance of the components used to explain differences in landscape fragility between several pilot areas. It was shown that all the components were significant (Recatalá 1995). The types defined for the last two landscape components are as follows:

1. Accessibility (AC).
    * AC1. Units with main road (motorway or highway).
    * AC2. Units with regional road.
    * AC3. Units with local road.
    * AC4. Units adjacent to AC1, AC2 or AC3.
    * AC5. Units without roads, not adjacent to AC1, AC2 or AC3.

2. Potential presence of observers (O).
    * O1. < 1000.
    * O2. Between 1000 and 5000.
    * O3. Between 5000 and 20000.
    * O4. Between 20000 and 50000.
    * O5. > 50000.

A weighting-rating method, similar to the one described above, was also used to assess intrinsic visual fragility of landscapes in a region. The values assigned to the different classes of the landscape components are also shown in Table 1. The weights obtained for the different landscape components are: relief-gradient = 2, relief-unevenness = 1, altitude = 1, lithology = 1, waterbodies = 2, land use = 1, vegetation = 2, human constructions = 2, accessibility = 2, and potential presence of observers = 2.

Intrinsic visual fragility is calculated using the following expression:

$$FVI_{pu} = \sum_{i=1}^{n} p_i \times vif_{pu}$$

where FVI_{pu} = intrinsic visual fragility of landscape p which is the visual image of mapping unit u, p_i = weight for landscape component i, n = number of landscape components, and vif_{ipu} = value of intrinsic visual fragility for landscape component i.

As for QVI values, FVI values have only a qualitative and relative meaning for comparison purposes.

2.3 Extrinsic visual quality

This property includes the consideration of the effect on the visual quality of a unit of those units surrounding it within the same visual basin. The idea is that the intrinsic value of a landscape unit is also a function of the quality of its surroundings, which may add to or detract from it.

The assessment of this property, therefore, requires the identification and representation of visual basins. Several techniques, both manual and computer-assisted are available for this purpose (Claver Farias et al. 1982).

The following expression is proposed for extrinsic visual quality:

$$QVE_{pu} = \frac{\sum_{i=1}^{n} QVI_{qi} \times S_i}{SCV_u - S_u}$$

where QVE_{pu} = extrinsic visual quality of landscape p which is the visual image of mapping unit u, QVI_{qi} = intrinsic visual quality of landscape q (visual image of mapping unit i) which is included in the visual basin of landscape p, n = number of landscapes included in the visual basin of landscape p, S_i = area of mapping unit i included in the visual basin of landscape p, SCV_u = area of the visual basin of landscape p, and S_u = area of mapping unit u.

The expression above takes into account the fact that the contribution of surrounding units depends on their relative area.

2.4 Extrinsic visual fragility

As in the case of quality, this property refers to the influence the intrinsic visual fragility of the surrounding landscape may have on a given unit.

A similar expression is proposed:

$$FVE_{pu} = \frac{\sum_{i=1}^{n} FVI_{qi} \times S_i}{SCV_u - S_u}$$

where FVE_{pu} = extrinsic visual fragility of landscape p which is the visual image of mapping unit u, FVI_{qi} = intrinsic visual fragility of landscape q (visual image of mapping unit i) which is included in the visual basin of landscape p, n = number of landscapes included in the visual basin of landscape p, S_i = area of mapping unit i included in the visual basin of landscape p, SCV_u = area of the visual basin of landscape p, and S_u = area of mapping unit u.

2.5 Landscape environmental quality

Landscape environmental quality is defined as the combination of visual quality and fragility. Accordingly to Cendrero et al. (1986, 1990) and AMA (1989, 1990), the differential contribution of each property to environmental quality is based on the fact that visual quality is more relevant than fragility for the perception of landscape value. A weighted sum is used for the integration:

$$EQ_{pu} = 2QVI_{pu} + 2QVE_{pu} + FVI_{pu} + FVE_{pu}$$

where EQ_{pu} = environmental quality of landscape p which is the visual image of mapping unit u, QVI_{pu} = intrinsic visual quality of landscape p, QVE_{pu} = extrinsic visual quality of landscape p, FVI_{pu} = intrinsic visual fragility of landscape p, and FVE_{pu} = extrinsic visual fragility of landscape p.

The weights in the previous formula indicate that visual quality (intrinsic and extrinsic) is twice as important than visual fragility (intrinsic and extrinsic) for the definition of landscape environmental quality. Obviously, the higher the EQ_{pu} value, the higher the environmental quality of landscape p is. As for the previous values (QVI, QVE, FVI and FVE values), EQ values have only a qualitative and relative meaning that allows comparisons between different landscapes in a region in terms of landscape merit. For comparison purposes it is useful to define a reference system with several classes of landscape merit. This can be done using the mean value (M) and standard deviation (s) of the values obtained in a region. Five classes have been defined:

Class I = Very high merit (EQ value > M + 1.25s).
Class II = High merit (EQ value between M + 0.5s and M + 1.25s).
Class III = Moderate merit (EQ value between M - 0.5s and M + 0.5s).
Class IV = Low merit (EQ value between M - 1.25s and M - 0.5s).
Class V = Very low merit (EQ value < M - 1.25s).

The assessment of extrinsic visual properties is time consuming when there are many mapping units in a study area. In the case of general land-use planning, it could be enough to assess landscape environmental quality on the basis of intrinsic visual properties only. In such case, landscape environmental quality could be expressed as:

$$EQ_{pu} = 2QVI_{pu} + FVI_{pu}$$

where EQ_{pu}, QVI_{pu}, and FVI_{pu} have the same meaning as in the previous expression.

Using this simpler approach, classes could be defined by dividing the difference between maximum ($LQmax$) and minimum ($LQmin$) values at equal intervals; d = (LQmax − LQmin)/n° of classes . This last value (d) refers to the equal contribution to each class of the classification reference system by the difference between $LQmax$ and $LQmin$. Specifically, the classes are the following:

Class I = Very high merit (EQ value between than EQ_{max} and EQ_{max} - d)
Class II = High merit (EQ value between EQ_{max} - d and EQ_{max} - $2d$)
Class III = Moderate merit (EQ value between EQ_{max} - $2d$ and EQ_{max} - $3d$)
Class IV = Low merit (EQ value between EQ_{min} + $2d$ and EQ_{min} + d)
Class V = Very low merit (EQ value between EQ_{min} + d and EQ_{min})

3 LANDSCAPE ENVIRONMENTAL QUALITY FOR CONSERVATION AND EIA PURPOSES

Assessment and classification of landscape environmental quality in a region provides useful information for establishing strategies for conservation purposes and for minimizing impacts. Such strategies should include the conservation of the most valuable areas in terms of landscape environmental quality (mapping units with landscapes of high and very high values).

The system proposed also provides the means to assess landscape impacts on a numerical basis. As shown in Table 1, the landscape components considered are basically of two types: a) "permanent" components which correspond to geomorphological characteristics (1-4 and, in most cases, 5); b) components which depend on or may vary through human activities (6-10 and, in some cases, 5). This means that the "types" of those components may be different before and after a given project is carried out and, therefore, the quality, fragility and environmental quality values would be different. If a numerical value (very high = 5; very low = 1) is assigned to each class, landscape impacts can be expressed as the difference between the post-project and pre-project situation. If the latter is higher, the impact would be negative and viceversa. The numerical value thus obtained can be used to rate the impact (-4 = critical; -3 = severe; -2 = moderate; -1 = low; 0 = very low-nil).

The assessment of impacts includes the consideration of a variety of characteristics, such as magnitude, importance, temporality, intensity, reversibility, and so on. The method proposed is particularly useful to establish importance (severity) and magnitude (extent of units affected), but it can also help to assess others, through the consideration of changes in individual landscape components.

4 LANDSCAPE ENVIRONMENTAL QUALITY FOR CONSERVATION AND EIA PURPOSES; A CASE STUDY

The methodology developed before was applied to a case study in the Sagunto area (Valencia) at the scale 1:25 000 (Recatalá & Sánchez 1996a, Recatalá et al. unpubl.b).

4.1 *Description of the study area*

The study area is located at the north eastern corner of the province of Valencia (Fig. 1). Sagunto (56600 inhabitants) is the main centre of the area, which has a total population of about 69000 people and 114 km^2. The region has a dynamic economy, dominated by industry, tourism and intensive, irrigated agriculture.

The region has a dry temperate Mediterranean climate characterized by a dry summer. Mean annual rainfall ranges from 450 to 550 millimetres. Mean annual temperature range from 15°C to 18 °C decreasing with distance from the sea. Climatic variability within the area results mainly from the influence of local topography and distance from the sea.

Two-thirds of the area (coastal plain) are formed by Quaternary sediments, including river alluvium, colluvium, coastal sands and lagoon deposits. Minor areas of Triassic, Jurassic and Tertiary materials occur inland; they constitute the hills and foothills of the Iberian chain.

Three major successive topographic zones run north-south roughly parallel to the coast. The western zone, which covers one quarter the area, has mountains up to 400 metres elevation. East of this zone there is a belt of hills rarely exceeding 250 metres, followed by a broad coastal lowland. The coastal plain encompasses alluvium deposits, as well as lagoons and sand dunes.

Quaternary sediments are associated with a great variety of soils. The extensive alluvium sediments present clay and fluvial soils. Sediments of lagoons support saline and/or hydromorphic soils. Dune sands have weakly developed shallow and sandy soils. Calcareous reliefs (limestones) with relatively gentle slopes form calcaric and chromic soils. Siliceous

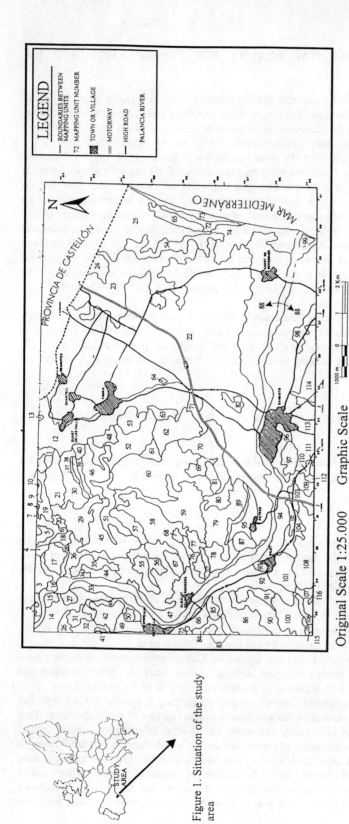

Figure 1. Situation of the study area

Figure 2. Mapping units of the study area.

reliefs (sandstones and claystones) are associated with poorly developed soils, and with sandy soils when the Bundsandstein facies occurs. Calcareous sediments in the hilly country support gravely and calcareous soils.

The potential climacic vegetation of the area corresponds to the class *Querceta ilicis* (Costa 1986). The main vegetation types which can be recognized are: brushwood pine trees, scrub combined with transitional communities and pastures associated to the latter. Non-climacic vegetation types include those specific of certain environments, such as lagoons or coastal dunes.

The main land uses are agriculture, forestry and industrial-urban uses. Irrigation crops are the predominant land use in the coastal zone, where intensive citrus cultivation takes place. Rain-fed agriculture is practiced inland even on quite steep slopes with soils subject to erosion. Industrial-urban uses are especially found along the coast, where numerous permanent and temporary houses and apartment buildings have been built in recent years. Residential cottages and forestry uses are found in the hilly and mountainous country.

These land uses are often in conflict with landscape conservation. Landscapes of lagoons and dunes in the coastal zone have been affected by irrigation crops and industrial-urban uses. Natural forest landscapes have been substantially affected by both the advance of rain-fed agriculture and uncontrolled forestry activities. The main threats for landscape conservation are probably the rapid residential and tourism development and forest fires.

4.2 *Application of the methodology*

Landscape assessment was carried out using as a basis the map of integrated environmental units at the scale 1:25 000 prepared by Diputación Provincial de Valencia, Universitat de València & Consejo Superior de Investigaciones Científicas (1984). The map (Fig. 2) has the form of a mosaic of units, "mapping units", which can be considered homogeneous from the point of view of the characteristics considered (Vink, 1963; Cendrero et al., 1976; Sánchez et al. 1984). A detailed description of these mapping units has been presented by Recatalá (1995). Data for the assessment of landscape quality in the units have been obtained from the former authors as well as Instituto Geológico Minero (1974), Dirección General de la Producción Agraria (1981), Prevasa (1982), Diputación Provincial de Valencia, Universitat de València & Consejo Superior de Investigaciones Científicas (1984), Servicio Cartográfico del Ejercito (1990) and Direcció General d'Urbanisme i Ordenació del Territori (1991).

The methodology described above was applied and the results obtained are shown in Figure 3 as well as Table 2.

Table 2. Area (%) belonging to each landscape environmental quality class

Classes	(%)
Class I (Very High)	0.82
Class II (High)	1.24
Class III (Moderate)	20.35
Class IV (Low)	75.57
Class V (Very low)	2.02

The map of landscape environmental quality shows that only one unit (on the hilly area with siliceous bedrock and well preserved vegetation) can be considered of very high merit. This results from very high intrinsic and extrinsic visual quality, and moderate intrinsic and extrinsic visual fragility.

A few mapping units in the fairly steep reliefs at the north of the study area correspond to the high environmental quality class. These units have high intrinsic and extrinsic visual quality, and moderate intrinsic and extrinsic visual fragility.

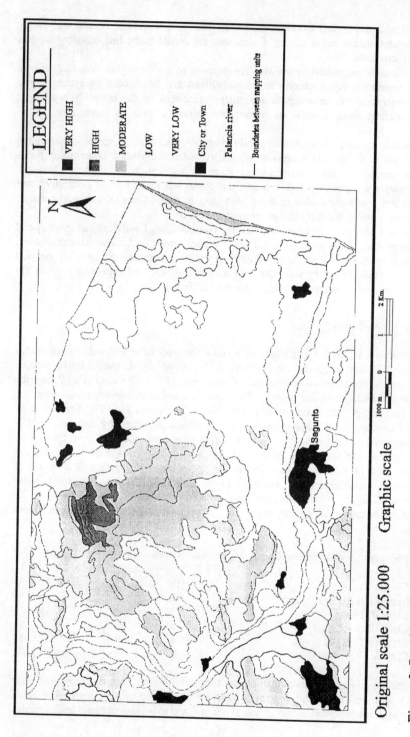

Figure 3. Landscape quality of mapping units in the study area.

Units of moderate landscape environmental quality are more abundant. They are located mainly in the hill country and on sand dunes. In general, they have moderate intrinsic and extrinsic quality and fragility, but other combinations may also occur.

The most abundant class, by far, is the low environmental quality one, which accounts for three quarters of the study area. Units of this class correspond essentially to alluvial terrains along the main river (Palancia) and the coastal plain.

Finally, a few units present very low landscape environmental quality; they are also on alluvial terrains, with high density construction or crops as dominant land uses.

4.3 *Discussion*

The results briefly presented above show that landscape value in the study area has experienced significant impacts in the past, as a consequence of human activities. These impacts are reflected in the decline in landscape "merit" (or landscape environmental quality) which affected mainly the hilly zone to the west of the study area and the coastal fringe. For instance, forest landscapes inland have been reduced as a result of the extension of agriculture, residential developments, excessive forest exploitation and/or reforestation.

The type of assessment described can be very useful for reducing environmental impacts when planning new activities in the area. Figure 3 shows that new activities located on the coastal plain, the fluvial valley and a significant part of the adjacent reliefs would affect units of low or very low landscape environmental quality and would, in general, be acceptable from this point of view. Moreover, when considering alternatives for the location of new activities, the method described would also enable to select the "least visual impact" solution between several units of the same landscape environmental quality; this solution would correspond to the unit with the lowest fragility.

The landscape environmental quality map of Figure 3 could obviously be useful for defining landscape conservation or rehabilitation policies and plans. Very high or high merit units should have priority for preservation; rehabilitation action should focus on moderate merit units adjacent to the former. This, naturally, would help to prevent the impacts of future activities.

The method also provides the means to express visual impacts in a numerical form. As described in the methodology, the quality, fragility and environmental quality of a landscape can be considered as the result of the combined perception of a series of heterogeneous features, mainly geomorphology, land cover and human activities. This combined perception is in the end presented in the form of a five-class hierarchy. If "point values" from 1 to 5 are assigned to the classes, landscape impacts can be expressed as the difference between landscape merit (or visual quality or visual fragility) after and before the implantation of the activity. For instance, unit No. 100 (Figs. 2, 3) has very high merit (5); if an urban or industrial use were to be implemented on this unit, landscape merit would decrease to low (2) or very low (1). Visual impact could thus be expressed as -3 (severe) or -4 (critical).

This approach can be particularly useful for selecting the least impacting solution among several alternative routes for linear structures such as roads, railways, power lines, etc.

5 CONCLUSIONS

The approach presented here helps to make the assessment of impacts on the landscape more systematic, transparent and objective. The geomorphological, land cover/land use and human variables used for the description of mapping units are expressed as point value ranks and integrated to obtain assessments of certain significant properties (intrinsic and extrinsic visual quality and fragility; merit) also in a numerical scale. These scales can then be used as a basis for "measuring" the visual impact of future actions or projects.

The results of the application of the methodology to the case study in the area of Sagunto show that landscape assessments thus carried our yield results which are reasonably coincident with the non-structured perception of the local population. These results indicate that over three quarters of the area have low or very low quality and could absorb most activities with an acceptable level of visual impact. The analysis has also allowed to define units which should have priority for conservation or for rehabilitation. The map presented represent a tool which can be particularly useful for selecting the least impacting solution among several alternatives, especially those concerning linear structures.

REFERENCES

AMA 1989. *Mapa Geocientífico de la Provincia de Castellón. Escala 1:200 000*. Valencia: Agencia del Medio Ambiente de la Generalitat Valenciana.

AMA 1991. *Mapa Geocientífico de la Provincia de Alicante. Escala 1:200 000*. Valencia: Agencia del Medio Ambiente de la Generalitat Valenciana.

Amir, S. & Gidalizon, E. 1990. Expert-based Method for the Evaluation of Visual Absorption Capacity of the Landscape. *J. Environ. Manage*. 30: 251-263.

Amir, S. & Sobol, A. 1987. *The use of geomorphological elements of the landscape as a basis for visual analysis of the Mediterranean Coast of Israel*. Haifa: Faculty of Architecture & Town Planning.

Arthur, L.M., Daniel, T.C. & Boster, R.S. 1977. Scenic Assessment: An Overview. *Landscape planning* 4:109-129.

Bolos, M., Bovet, T., Estruch, X., Pena, R., Ribas, J. & Soler, J. 1992. *Manual de Ciencia del Paisaje. Teoría, métodos y aplicaciones*. Barcelona:Masson.

Boluda, R., Molina, M.J. & Sánchez, J. 1984. Definición y metodología de unidad de paisaje. Importancia de la Geología Ambiental en su definición. *Comunicaciones del I Congreso Español de Geología*: 611-622. Segovia:Colegio Oficial de Geólogos.

Carls, E.G. 1974. The effects of people and man-induced conditions on preferences for outdoor recreation landscapes. *J.Leisure Res.* 2:113-124.

Carlson, A.A. 1977. On the possibility of quantifying scenic beauty. *Lanscape Planning* 4:131-172.

Cendrero, A., Díaz de Terán, J.R. & Saíz de Omeñaca, J. 1976. A technique for the definition of environmental geologic units and for evaluating their environmental value. *Landscape Planning* 3: 35-66.

Cendrero, A., Nieto, M., Robles, F., Sánchez, J., Díaz de Terán, J.R., Francés,E.,González-Lastra, J.R:, Boluda, R., Garay, P., Gutiérrez, G., Jiménez, J., Martínez, J., Molina, M.J., Obarti, J., Pérez, A., Pons, V., Santoyo, A. & Stübing, G. 1986. *Mapa Gecientífico de la Provincia de Valencia*. Valencia: Diputación Provincial de Valencia.

Cendrero, A., Sánchez, J., Antolín, C., Arnal, S., Díaz de Terán, J.R., Francés, E., Martínez, V., Moñino, M., Nieto, M., Nogales, I., Pérez, E., Rios, C., Robles, F., Romero, A. & Suárez, C. 1990. Geoscientific maps for planning in semi-arid regions: Valencia and Gran Canaria, Spain. *Eng. Geol.* 29: 291-319.

Cerny, J.W. 1974. *Scenic analysis and assessment*. CRC Critical Reviews in Environmental Control. Durham: University of New Hampsshire.

Claver Farias, I., Aguilo, M., Aramburu, M.P., Ayuso, E., Blanco, A., Calatayud, T., Ceñal, M.A.,Cifuentes, P., Escribano, R., Francés, E., Glaría, G., González, S., Lacoma, E., Muñoz, C., Otero, I., Ramos, A. & Saíz de Omeñaca, M.G. 1982. *Guía para la elaboración de estudios del medio físico: contenidos y metodología*.Madrid: Servicio de Publicaciones del Ministerio de Obras Públicas y Urbanismo.

Costa, M. 1986. *La Vegetación en el País Valenciano*. Valencia:Servicio de Publicaciones de la Universitat de València.

Díaz de Terán, J.R. 1985. Estudio geológico-ambiental de la franja costera Junquera Castro Urdiales (Cantabria) y establecimiento de bases para su ordenación territorial. PhD Thesis.Oviedo: Universidad de Oviedo.

Diputación Provincial de Valencia, Universitat de València & Consejo Supeior de Investigaciones Cientificas 1984. *Los suelos de la provincia de Valencia: su evaluación como recurso natural.* Hoja de Sagunto (668-II). Valencia: Diputación Provincial de Valencia.

Dirección General de la Producción Agraria 1981. *Mapa de cultivos y aprovechamientos. Escala 1:50.000. Sagunto.* Madrid:Servicio de Publicaciones Agrarias del Ministerio de Agricultura, Pesca y Alimentación.

Direcció General d'Urbanisme i Ordenació del Territori 1991. *Plà Director d'Ordenació del Territori a la Comunitat Valenciana.* Valencia:Conselleria d'Obres Públiques i Urbanisme de la Generalitat Valenciana.

Dunn, M.C. 1974. *Landscape evaluation techniques: An appraisal and review of the literature.* Birmingham: Centre for Urban and Regional Studies of the University of Birmingham.

Escribano, M. 1987. *El Paisaje.* Madrid:Publicaciones de la Dirección General de Medio Ambiente del Ministerio de Obras Públicas y Urbanismo.

Fines, K.D. 1968. Landscape Evaluation: A research project in East Sussex. *Regional Studies* 2: 41-55.

Garling, T. 1976. The structural analyses of environmental perception and cognition. A multidimensional scaling approach. *Environment and Behavior* 8: 385-415.

Gómez Orea, D. 1978. *El medio físico y la planificación.* Madrid: CIFCA.

González-Alonso, S., Aramburu, P. & Garcia-Abril, A. 1995. Using the inventory: models in lanscape planning. In E. Martínez-Falero & S. González-Alonso (eds), *Quantitative Techniques in Landscape Planning*: 27-46. Florida: CRC Press.

González Bernáldez, F. 1981. *Ecología y Paisaje.* Madrid:Blume.

Ibáñez Martí, J.J., García Alvarez, A. & Monturiol Rodriguez, F. 1987. Ecología del paisaje: propuesta de una metodología para la prospección del medio físico en áreas de montaña mediterránea. *Actas de la III Reunión Nacional de Geología Ambiental y Ordenación del Territorio*: 1067-1084. Valencia: Sociedad Española de Geología Ambiental y Ordenación del Territorio.

Instituto Geológico-Minero 1974. *Mapa Geologico de España. E. 1:50.000. Sagunto (668).* Madrid:Servicio de Publicaciones del Ministerio de Industria y Energía.

Kaplan, S. & Kaplan, R. 1978. *Humanscape: environment for people.* Belmont:Duxbury Press.

Litton, R.B. & Tetlow, R.J. 1974. *Water and landscape: an aesthetic overview of the role of water in the landscape.* New York:Water Information Centre.

Macia, A. 1979. Factores de personalidad y preferencias en la eleccion de paisajes. PhD Thesis. Madrid: Universidad Autónoma de Madrid.

Prevasa 1982. *Situacion actual, problemas y perspectivas de las comarcas valencianas. IX. El Camp de Morvedre. Estudios basicos para la ordenacion del territorio de la Comunidad Valenciana.* Valencia:Prevasa-Caja de Ahorros de Valencia.

Ramos, A., Aramburu, M.P., Ayuso, E., Blanco, A., Ceñal, M.A., Cifuentes, P., Escribano, R., Glaria, G., González, S., Mantilla, P., Muñoz, C., Otero, I., & Saíz de Omeñaca, M.G. 1979. *Planificación física y ecológica. Modelos y Métodos.* Madrid: Magisterio Español, S.A.

Ramos, A., Ramos, F., Cifuentes, P. & Fernandez-Cañadas, M. 1976. Visual landscape evaluation, a grid technique. *Landscape Planning* 3: 67-88.

Recatalá, L. 1995. *Propuesta Metodologica para Planificacion de los Usos del Territorio y Evaluacion de Impacto Ambiental en el Ambito Mediterraneo Valenciano.* PhD thesis. Valencia:Servicio de Publicaciones de la Universitat de València.

Recatalá, L. & Sánchez, J. 1996a. Aplicación de la metodología de evaluación de la calidad ambiental del paisaje en el ámbito mediterráneo al cuadrante II a escala 1:25 000 de Sagunto. In J. Chacón & C. Irigaray (eds), *Riesgos Naturales, Ordenación del Territorio y Medio Ambiente*: 153-165. Granada: Sociedad Española de Geología Ambiental y Ordenación del Territorio.

Recatalá, L. & Sánchez, J. 1996b. Metodología de evaluación de la calidad ambiental del paisaje para planificación de los usos del territorio y evaluación de impacto ambiental en el ámbito mediterráneo. In J. Chacón & C. Irigaray (eds), *Riesgos Naturales, Ordenación del Territorio y*

Medio Ambiente: 137-151. Granada: Sociedad Española de Geología Ambiental y Ordenación del Territorio.

Recatalá, L., Ive, J.R. & Sánchez, J. Unpubl.a. Evaluation and classification of resouce units quality for conservation and environmental impact assessment. 1. A multicriteria approach.

Recatalá, L., Sánchez, J. & Ive, J.R. Unpubl.b. Evaluation and classification of resource units quality for conservation and environmental impact assessment. 2. A case study.

Sánchez, J., Arnal, S., Antolín, C. & Rubio, J. 1984. Metodología de la Cartografía Básica. *Actas del I Congreso Nacional de Ciencia del Suelo:* 771-782. Segovia:Servicio de Publicaciones del Ministerio de Obras Públicas y Urbanismo.

Servicio Cartográfico del Ejercito 1990. *Mapa topográfico general serie L de Sagunto. Hoja no.29-26 (668). Escala 1:50 000.* Madrid: Servicio de Cartografia Militar de España del Ministerio de Defensa.

Vink, A.P.A. 1963. Aerial photographs and the soil sciences. UNESCO. París.

Villarino, M.T. 1985. El Paisaje. *Curso sobre evaluaciones de Impacto ambiental.* Madrid: Servicio de Publicaciones de la Dirección General del Medio Ambiente del Ministerio de Obras Públicas y Urbanismo.

Zube, E.H. 1974. Cross-disciplinary and intermode agreement on the description and evaluation of landscape resources. *Environmental Behaviour* 6(1): 69-89.

Zube, E.H. 1984. Themes in landscape assessment theory. *Landscape Journal* 3: 104-110.

Westman, W.E. 1985. *Ecology, Impact Assessment and Environmental Planning.* New York:Wiley Interscience.

GIS applications for Environmental Impact Assessment. An analysis of the effects of a motor way project in a alpine valley

A. Patrono
International Institute for Aerospace Survey and Earth Sciences (ITC), the Netherlands

F. de Francesch
Servizio Geologico, Provincia Autonoma di Trento, Italy

M. Marchetti
Dipartimento di Scienze della Terra, Università degli Studi di Modena, Italy

A. Moltrer
Ufficio Valutazione di Impatto Ambientale, Provincia Autonoma di Trento, Italy

ABSTRACT: This contribution introduces GIS applications and geomorphologic characterisation as important elements for making environmental impact assessment operational. A motorway project is analysed which is located in a typical alpine environment near the city of Trento, in northern Italy. The plan considers a series of feasible solutions to expand the connection between two major road networks that cross a series of ecosystems where geomorphologic factors play a fundamental key role. The impacts are evaluated and converted to a computational form, suitable for spatial data analysis, multi-criteria analysis and decision-making by alternative ranking. Both the strategy of approach and the results of the analysis are discussed at the end of the contribution.

1 INTRODUCTION

Environmental impacts can vary in directness, intensity, and duration depending upon the nature of the human action and of the affected biotic-abiotic communities. In recent years a number of studies have focused on the role of spatial information systems in landscape planning and environmental impact assessment (EIA) with special attention to the potential of geographical information systems, or GIS, in environmental analysis (Scholten & Stillwell 1990, Carver 1991, Nijkamp & Scholten 1993, Pereira & Duckstein 1993, Patrono 1994, Fabbri et al. 1995, Patrono et al. 1995). With an actual application, a motorway project, this study makes GIS / EIA analyses effective and critical for decision-making, emphasising and focusing on the role of geomorphology, geology and land-use / land-cover in the assessment of the impacts.

2 THE PROJECT

The study area which approximately covers 40 Km^2 in the Province of Trento, in northern Italy, is shown in Figure 1. It includes part of the Val (valley) di Non and part of the Val d'Adige. The area is characterised by Alpine geomorphology and strong relief energy. The altitude ranges between 200 m above sea level at the bottom of the Val d'Adige to 2000 m above sea level in correspondence of the surrounding peaks.
  The purpose of the project analysed in this contribution is to develop a road connection between the main road network of the Val d'Adige and the one of the Val di Non and Val di Sole (a little further North), both characterised by strong economy, based on agriculture and tourism. The main goal of this study is to perform that part of the EIA analysis strictly related to the impacts caused by the proposed alternatives on the geomorphologic, geologic and land-use / land-cover resources of the area, which heavily characterise the considered environment.
  Nowadays, the mentioned connection is a narrow road passing through the town of Mezzolombardo and continuing towards the Val di Non with a winding route that causes a strong resistance to the traffic both outside and within the town centre.

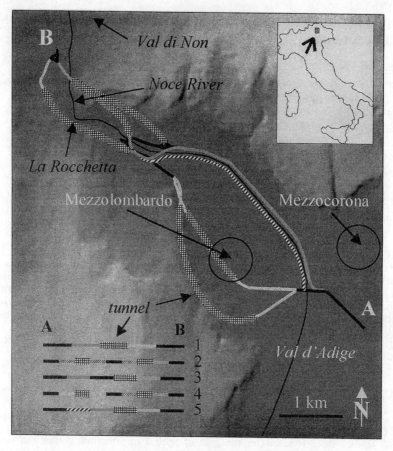

Figure 1. Geographical setting of the study area and layouts of the five proposed alternatives (A-B represent the two tie points to be connected).

The criteria considered by the team of planners charged by the Province of Trento in selecting the candidate alternatives aimed mainly at planning a road assuring a fast traffic, minimising the consumption of precious agricultural land and, in general, minimising the impact on residential areas. According to measured traffic flows, a "two lines - road solution" was adopted (average width of 15 m). The starting point, A in Figure 1, of the new road is assumed to be the same for all the alternatives and it coincides with the turn-off at Mezzocorona from the highway crossing the Val d'Adige. The end-of-road connection is inserted into the main new road (now under construction) crossing the Val di Non, at point B in Figure 1, placed on the right side of the Noce River. The length of the existing road from the starting point to the end of the road connection is approximately 7 km.

As a result of a feasibility study by the charged team, five alternatives had been proposed, which are also indicated in Figure 1. The technical parameters are specified in Table 1. They can be described as follows.

Table 1. Technical parameters of the alternatives.

Technical parameters	ALT. 1	ALT. 2	ALT. 3	ALT. 4	ALT. 5
Total length	6800 m	6150 m	5650 m	6850 m	6270 m
Total length of "open" ways	4260 m	2700 m	3700 m	2580 m	4200 m
Total length of tunnels	1700 m	3100 m	1700 m	3800 m	1700 m
Total length of bridges and viaducts	840 m	350 m	250 m	470 m	370 m

Alternative 1: the first part runs along the left bank of the Noce River and crosses it immediately before the gorge of "La Rocchetta", at the beginning of the Val di Non. It

continues with a tunnel of about 1700 m in length in the West side of the valley. It opens and links up with the main new road of the Val di Non through a viaduct that crosses a swampy area.

Alternative 2: it consists of building an artificial tunnel under the main road crossing Mezzolombardo. A tunnel inside the east side of the Val di Non bypasses "La Rocchetta". At the end, the tunnel is linked with the main road on the left side of the Noce River through a viaduct.

Alternative 3: the first part of the project follows the track of alternative 1. In the proximity of the "La Rocchetta" a tunnel starts on the East side of the valley. The tunnel opens upstream of the gorge and links up with the main road through a viaduct as above in Alternative 2.

Alternative 4: the alternative first follows the existing road as far as the entry into Mezzolombardo where a tunnel starts, which by-passes the town for about 2500 m in length and continues, after crossing "La Rocchetta", with another tunnel similarly to alternative 2.

Alternative 5: it is slightly different from the first solution only in the starting part because it is planned along the right side of the Noce River. It avoids to built the viaduct across the river and it permits to use the existing bridge.

3 ENVIRONMENTAL DESCRIPTION OF THE STUDY AREA

Evaluating environmental settings for EIA involves making measurements for a series of indicators and then deciding which areas are most significant based on those measurements. This implies to define for each area a certain value of environmental quality or importance. The selection of the indicators aims at enhancing or detracting the suitability of a proposed action (e.g. the set of alternatives), whose effects are commonly known as impacts. In this contribution, the purpose is to identify "the most valuable areas" so that the decision-making process aims at the best compromise in maintaining the areas' values. The indicators or factors of importance here considered are: geomorphology, geology and land-use / land-cover. This combination was selected because of the intricate system of processes and relations that operate at many different spatial and temporal scales within the three indicators.

3.1 *Geology*

In the area of interest a sedimentary sequence is present which is dated between the Upper Permian and the Palaeocene and which covers the basement of volcanic rocks termed "Piattaforma porfirica atesina" (porphyry platform). According to Prosser and Selli (1991), the formations characterising the area may be grouped into several units that show different competence and mechanical behaviour during deformation. The description is also referring to the geological map of the area in Figure 2 and to Table 2, including the legend specifications. The map is derived from geological map of Geological Survey of the Province of Trento (Provincia Autonoma di Trento, unpubl., see also Bartolomei et al. 1969). The first unit relevant for the study includes a massive sequence of peritidal limestones and dolomites about 1 km thick, consisting of the Dolomia della Val d'Adige Formation (Ladinian, 500 m thick), and of the Dolomia Principale Formation (Norian-Rhaetian, 350 m thick). The Calcare di Torra Formation (Lias) is almost absent in the Val d'Adige, North of Mezzocorona. In the proximity of "La Rocchetta", it crops out with typical faces and thickness (about 300 m). The second unit includes the Jurassic, Cretaceous and Palaeogene sequence consisting of thin bedded limestones and marls (Rosso Ammonitico, Biancone, Scaglia Rossa, Scaglia Grigia Formations). They represent the drowning of the Upper Triassic or Lower Jurassic platform followed by the onset of pelagic, often condensed sedimentation. The total thickness is about 300 - 400 m and the Eocene grey marls outcrop only in the lowermost Val di Non.

Focusing on the project, in the western sector, where there are the road layouts (compare Figures 1 and 2), calcareous-dolomitic rocks in thick strata and calcareous marly rocks (Ladinian - Upper Cretaceous) are exposed. At the entry of the Val di Non, connected with the hinge of homonymous syncline, the marly limestones and the Cretaceous - Palaeocenic marls are encountered.

The valley bottom is occupied by recent alluvium due to the sedimentation of the Noce and Adige Rivers. The sediments forming the fan of the Noce River consist of alternations of gravel and sands. The Adige River fluvial deposits are finer than those of the Noce River domain. Between the two types of deposits a zone of transition is located upstream of the "La Rocchetta" gorge.

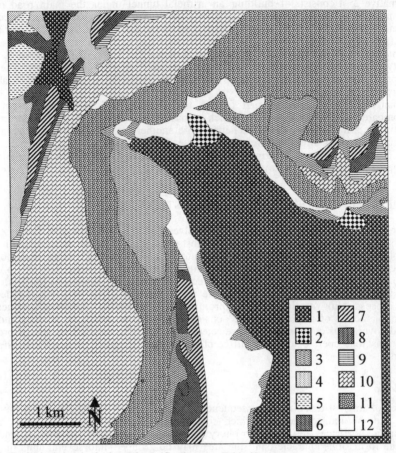

Figure 2. Geological sketch (for legend specifications see Table 2).

Table 2. Geological classes, O.V.: Original class value (Fig. 2); R.V.: Reclassified value (Fig. 5b).

O.V.	Geological classes	Material	R.V.
1	Holoc. alluvial deposits (Quaternary)	Sand and gravel	1
2	Alluvial fan (Quaternary)	Sand and gravel	1
3	Debris (Quaternary)	Coarse gravel	1
4	Moraine (Quaternary)	Gravel and sand in silty matrix	1
5	Old alluvial deposits (Quaternary)	Conglomerate	1
6	Scaglia grigia (Palaeocene - Eocene)	marls	2
7	Scaglia rossa (upper Cretaceous)	marly limestone	2
8	Biancone (upper Cretaceous - Malm)	Stratified marly limestone	2
9	Rosso ammonitico (Malm - Dogger)	Nodular limestone	2
10	Calcare di Torra (Lias)	Grey micritic limestone in thick strata	3
11	Dolomia Principale (Norian-Rhaetian)	Dolomite	3
12	Dolomia Val d'Adige (lower Ladinian)	Dolomite	3

Moraine, old alluvial deposits and debris represent the quaternary cover of the slopes. The moraine deposits are mixed material both in composition and granulometry, and mostly consist of pebbles and gravel in a sandy and silty matrix. In the lowest part of the Val di Non, old alluvial deposits, which mostly consist of, cemented breccias locally outcrop. At the foot of the steeper rocky surfaces there are debris forming cones mostly made up of

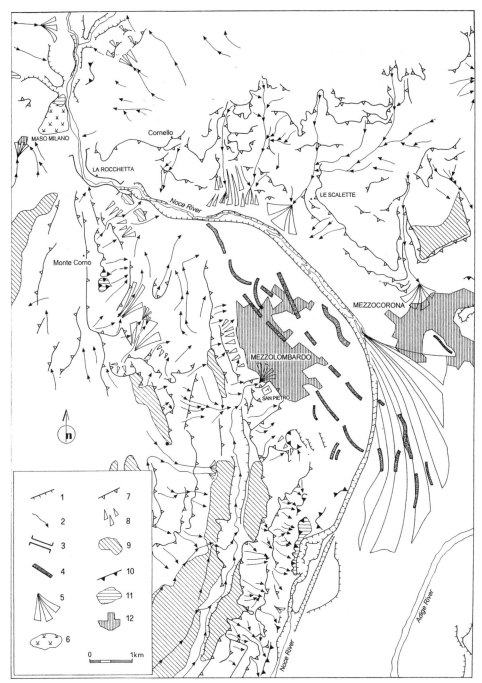

Figure 3. Geomorphologic map: (1) Fluvial scarp, (2) Linear erosion, (3) Fluvial gorge, (4) Paleochannel, (5) Alluvial fan, (6) Swampy area, (7) Scarp mainly due to slope processes, (8) debris, (9) Glacial areas (10) Scarp due to human activities, (11) Quarry area, (12) Built up area.

material of intermediate and small grain size with sharp edges, indicating progressive crumbling because of thermoclastism connected to the climatic conditions at the closing of the last glacial cycle.

The geological structure of the area is typical of a regional tectonic complex system with a predominant NorthWest - SouthEast orientation, represented by NorthEast - SouthWest thrusts and related subvertical transfer faults, connected to an orogenic event older than the Alpine one. The structural system causes a diffuse jointing of the most massive rocks oriented along the main deforming systems. The marly stratified lithotypes have induced a plastic deformation recognisable overall at the entrance of the Val di Non where the hinge of a syncline is present.

3.2 *Geomorphology*

Geomorphologically, the study area is characterised by the fluvial deposits of the Noce River from the narrow valley North of "La Rocchetta" to the Val d'Adige. The area represents an ideal setting for vines, from the geopedological point of view. Among others, the well known "Teroldego Rotaliano" wine is produced in the vineyards of the valley bottom and it is classified as a "DOC" (Denominazione di Origine Controllata) and very high quality product, typical of the area.

The plain ("Piana Rotaliana") is characterised by several abandoned fluvial courses, which have been almost completely erased by the farming activity. It is possible to distinguish the boundary however between the Noce River alluvial fan and the Adige River fluvial deposits. Human activities and properties are concentrated in the alluvial plain while the steep slopes are still in a natural state. The slopes, especially the Dolomitic ones, show scarps that present high relief energy which in the high elevations show jumps of about 600 meters. The scarps North and North-West of Mezzocorona, North-West of Mezzolombardo and at "La Rocchetta" gorge are particularly significant also from the aesthetic point of view being large and attractive. The slopes are characterised by gullies, which do not appear to affect slope stability. At the bottom of the slopes, deposits occur principally due to gravity, which conceal the boundary between bedrock and fluvial deposits. They are often used as row material. Above the Dolomitic scarps a plateau characterised by low relief energy dominates the landscape. The plateau consists of less resistant lithologies and also of landforms and deposits due to the glacial morphogenesis or to fluvioglacial or fluvial phases, which are older than the last glaciation. The Val di Non upstream of the "La Rocchetta" gorge shows a swampy area that is of great ecological interest because it is a very rare and particular biotope at regional level (Peer 1990). Upstream several scarps can be seen to bound fluvial terraces. These terraces are due to the Noce River and testify its incrused erosion after the deepening of the river-bed at "La Rocchetta" gorge. Figure 3 shows the geomorphologic sketch of the area, derived by aerial photo interpretation and implemented with the geomorphologic map of the Province of Trento (Comprensorio Valle dell'Adige, unpubl.).

Table 3. Geomorphologic classes.

CL.V.	Geomorphologic class
1	Area with alluvial deposits
2	Alluvial fan
3	Area with slope debris
4	Fluvial gorge
5	Area with glacial deposits
6	Man-modified area
7	Area with fractured rocks
8	Swampy areas
9	Bedrock
10	Slope scarps

3.3 *Land-use / land-cover*

The natural vegetation cover in the study area can be subdivided in units characterised by the presence of main wooded species. Their distribution depends on the elevation range, slope, aspect and edaphic condition (Landolt 1977). Woods of *Ostrya carpinifolia* Scop., *Fraxinus ornus* L. and *Quercus pubescens* Willd, represent the warmer habitats. At higher elevations, in pioneer conditions, *Pinus nigra* Arnold and *Pinus sylvestris* L. are the representative species occurring, while *Fagus sylvatica* L. is typical of more favourable environments. The highest elevations are characterised by the presence of *Picea excelsa* (Lam.) Link and *Larix decidua* Miller.

Table 4. Land-use / land-cover classes. O.V.: Original value (Fig. 4); R.V.: Reclassified value (Fig. 5c).

O.V.	LAND-USE/-COVER class	R.V.
1	Vineyards	1
2	Apple trees	2
3	Vineyards - apple trees (mixed)	3
4	Evergreen vegetation	4
5	Beech woods - evergreen vegetation	4
6	Oak woods - evergreen vegetation	4
7	Oak woods	4
8	Beech woods	4
9	Grasslands	5
10	Rivers	6
11	Settlements	10
12	Bare rocks	10

The wooded cultivated areas are almost exclusively consisting of vineyards and some apple orchards. Their distributions generally cover the alluvial deposits at the valley bottoms and, eventually in terraces, at the lower or warmer parts of the slopes. The arable land has nowadays almost disappeared. The cultivation of apple trees and vines represents a considerable income for the entire area (Peer 1990).

Especially focusing on the area directly affected by the project, the presence of vineyards results critical. Almost the entire available land is planted with very specific cultivars and is managed by local wineries. The economic relevance of the wine business gives landowners a strong influence in the decision-making process. Comparing Figure 1 with Figure 4 it is evident that the acceptable solutions, resulting after the first step of feasibility analysis, were carefully selected and planned to preserve the present distribution of vineyards.

The land-use / land-cover map used in this contribution and shown in Figure 4, is part of the result of a pilot project also carried out in Trento for the provincial government (Patrono 1996). The main goals were (a) to evaluate the accuracy of the information obtained from Landsat TM for distinguishing and mapping different land-use / land-cover types in a heterogeneous sample area, and (b) to investigate the opportunities for using higher resolution sensors (like Spot and scanned aerial photos) in the classification process. The result was the definition of a method for combining images of different resolutions in order to improve automatically the scale fitting of the classified images with the topographic map (1:10,000) of the Province of Trento. The result of one of the supervised classifications is shown in Figure 4 and in Table 4 the class specifications are listed. (fig. 4, see color plate pages, following p. 225)

4 TMU ANALYSIS FOR EIA

In EIA, objective and subjective criteria have to be evaluated performing measurements which imply the quantification of impacts whenever possible. Such quantification is difficult for parameters describing the environment especially in the context of spatiality and time limitations. In addition, normalisation procedures, fundamental steps in EIA studies, require comparisons between different mapping units where significance, spatial representation, and distribution of indicators can vary (Patrono 1994; Patrono et al. 1995). Starting from a geomorphologic perspective, several methods were developed for assessing environmental impacts (Marchetti et al. 1995). Rivas et al. (1995) proposed a strategy to incorporate in EIA the impact on geomorphologic resources. This has been modified by Panizza et al. (1995) and Patrono et al. (1995) so that it could be applied using a GIS and its spatial modelling functions, to automate the measurement of objective criteria, and to establish at a general procedure.

In the analysis of a project, a series of phases have to be considered to deal with the introduction of environmental effects into the decision process. According to Beinat (1995) the three principal phases in which EIA plays a key role are: (a) planning of the potential actions and identification of the environmental sectors involved, (b) analysis of impacts and data organisation for the evaluation phase to follow, and (c) evaluation and ranking of alternatives. Thus, since the beginning, EIA deals with the decision process, which leads to the selection of the alternatives. In this particular study, risk, hazard and feasibility analysis have already been performed by the mentioned team of planners, so that the alternatives to be considered were all feasible. The environmental sectors here analysed are geomorphology, geology and land-use / land-cover and the main target is the evaluation of the impact of the alternatives on specific combinations of landforms, geological units and classes of land-use / land-cover.

In performing EIA it is first necessary to spatially identify operational terrain mapping units or TMUs. In general TMUs represent the geographic location of entities that have a unique sets of attributes (Meijerink 1988; van Westen 1993). From an EIA perspective, whenever a proposed action interests a TMU, it identifies a homogeneous landscape patch potentially subjected to the modifications caused by the impact(s). In this study a terrain mapping unit groups unique combinations of the three indicators selected for the analysis.

In defining the TMU map, the indicators had first been reclassified. A clustering is often required to limit the amount of all possible class combinations that generally result in an exaggerated fragmentation of the study area. This, in general, leads also to the presence of many, poorly defined and small polygons contained at the edges of potentially acceptable TMUs. Due to their similar characteristics, coherently with the performed analysis, a compromised situation was generated in which classes were merged as shown in Table 2 for the geology, Table 3 for the geomorphology and Table 4 for the land-use / land-cover. Figure 5 shows the consequently reclassified maps of the three indicators. (fig. 5, see color plate pages, following p. 225)

The geological formations occurring in the study area were grouped in units having similar morphogenetic and/or lithological characters and similar behaviour in relation to present geomorphic processes. In the identification of the grouping rules, tectonism, structural features and hydrogeological conditions were considered. The following three broad classes were identified (Tab. 2 and Fig. 5b):

(R.V. 1) Unconsolidated rocks. This class includes the Quaternary alluvial fan and deposits, debris, moraine, and old alluvial deposits. In particular the road layouts affect the alluvial bottom deposits which are generally stable with variable cohesion and angle of friction, and which have high bearing capacity, with decreasing value in the presence of water. The alluvial deposits are generally kept coherent by the vegetation or by the agriculture practice. They have a high permeability and the water table is connected with underdrained bed, which has seasonal variations and is exploited for drinking water and irrigation.

(R.V. 2) Rigid-plastic consolidated rocks. This class includes: Scaglia grigia, Scaglia rossa, Biancone, and Rosso ammonitico Formations. This group includes rocks consisting of beds with alternation levels of plastic rocks, marls and more rigid rocks (mostly limestone) with variable thickness. Structural elements such as faults, overthrusts, joints, and bedding planes are critical in determining the slope stability of rocks mass and the underground water circulation as well.

(R.V. 3) Massive calcareous consolidated rocks. This class includes: Calcare di Torra, Dolomia Principale, Dolomia Val d'Adige Formations. They are characterised by a complex stratigraphy or occur in thick strata. The rocky mass is affected by disjunctive elements (faults, joints) which make it a heterogeneous and discontinuous medium. The rocks in this class have scarps that are several hundreds of meters high. The relative solubility of the lithotypes causes a progressive widening of joint and consequently leads to karstic water circulation.

The geomorphologic sketch shown in Figure 3, is not directly usable for digital integration with the geological and land-use / land-cover maps. For this reason a derived map (Tab. 3) was prepared from the original one and it is shown in Figure 5a. The criteria used to prepare the new map are as follows: (a) the bedrock is always divided by deposits; (b) deposits are subdivided in: (1) alluvial deposits in the floodplain; (2) alluvial-fan deposits and (3) slope deposits (scree-slope and talus cone); (c) some important geomorphologic elements of the landscape are extracted as pre-defined polygons: (4) fluvial gorges, (5) glaciated areas, (6) man-modified areas (built up area, quarry area, etc.), (7) fractured rocks, (8) swampy areas, (9) bedrock and (10) scarps.

The land-use / land-cover classes were reclassified considering the following criteria: (a) the wooded cultivation, vineyards, apple orchards and mixed situations, were maintained as separated classes because of their particular economic importance (i.e. reclassified as classes (1), (2) and (3) respectively as shown in Table 4); (b) the natural vegetation had been reclassified in two structurally homogeneous groups, which are woods and grasslands (respectively classes (4) and (5) in Table 4); (c) due to their specific nature, classes "rivers", "settlements", and "bare rocks" were unmodified and reclassified as classes (6), (10) and (10), as in Table 4 (10 is the "irrelevant" value).

Table 5. Data of the unique-condition areas obtained by a query for the criteria ("Cr") 2(1) and 5 (Tab. 6). "Key": area identifier. "Cells": number of cells of the area. "Lg, Lo, Lp, Le, C and Q.": see expression (1). "Imp. 1, 2, 3, 4 and 5": number of cells impacted by Alt. 1, 2, 3, 4 and 5.

Key	Cells	Cr	Lg	Lo	Lp	Le	C	Q	Imp. 1	Imp. 2	Imp. 3	Imp. 4	Imp. 5
25	315	21	0	0	0	0.75	1	0.75	48	0	48	0	0
39	6	21	0	0	0	0.75	1	0.75	0	0	0	0	0
41	58	21	0	0	0	0.75	1	0.75	0	0	0	0	0
48	5451	21	0.25	0	0.5	0.75	1	1.5	555	464	555	464	464
49	13	21	0.25	0	0.5	0.75	1	1.5	0	0	0	0	13
66	12	21	0	0	0	0.75	1	0.75	0	0	0	0	0
03	67	5	0	0.25	0	0	0.75	0.1875	0	0	0	0	0
11	160	5	0.25	0.25	0	0	1	0.5	14	40	40	40	14
22	2090	5	0.25	0.25	0	0.5	0.75	0.75	453	36	458	36	372
76	298	5	0	0	0	0.5	1	0.5	0	0	0	0	0

Once the three indicator maps had been reclassified, the three maps were combined by digital overlaying also termed "crossing" in which a table is generated which identifies and characterises all the unique combinations of values. The resulting map is showing the pattern distribution of all possible combinations of the values in the three reclassified maps.

Using an area-numbering procedure, each of the subarea with unique conditions was assigned a unique sequential identifier. In this way it was possible to build up a relational data base and to associate with each single polygon of the cross map any kind of information that can be derived, with a "query", from the data base. The data base has been developed using the built-in table manipulation functions of ILWIS (ITC 1996), the GIS PC-based software used in this study.

The database characterises uniquely the spatial distribution of the indicators used in this study. The next step in the analysis was to extract only the areas needed in the environmental impact assessment. Table 5 shows part of the database. The criteria used for the selection of the subareas with different indicator values are summarised in Table 6. The criteria, expressed as combinations of indicators, were selected to highlight the critical links between them in a geomorphologic perspective. In Table 6, for each criterion, the leading indicator is outlined in italics. The geomorphology does not necessarily represent the critical key for the selection. As mentioned in the description of the three indicators, it is the relationship with the geomorphologic setting that characterises the selection of criteria and that also represents the focus of the environmental study here analysed.

Due to its economic implication, for the first two combinations of indicators in Table 6, the presence / absence of the "DOC areas" was also considered. As previously mentioned, the DOC area identifies the production of wines whose quality and origin are controlled and guaranteed. Thus the impacts on the first two combinations of indicator values were considered differently if they occurred within or outside DOC areas. Figure 6 shows the unique-condition areas filtered from noise and obtained by a query for the combinations in Table 6. The image with the relevant indicator value areas was later used in the EIA.

5 IMPACT ASSESSMENT

The measurement of the impacts caused by the emplacement of an infrastructure on the natural environment in a given area can be seen as the task of multi-criteria analysis. Using the predictions of the potential impacts obtained from the analysis of the spatial database using a GIS, the decision-making process consists of ranking and selecting the alternatives (Janssen 1992). In the present contribution it was based on the comparison of the geomorphologic, geologic and land-use / land-cover importance of the areas affected by the different impacts. In the decision-making process, the higher is the "loss" of natural value

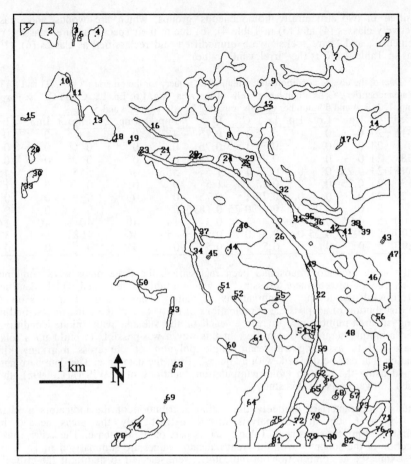

Figure 6. TMUs map with unique – condition areas obtained by a query for the combinations in Table 6. The number refers to the column "Key" in Table 5.

Table 6. The criteria are expressed as combinations of indicators. For each criterion, the leading indicator is highlighted in italics.

Criterion	Geology	Geomorphology	Land-use
1(1)* and 1(2)*	Unconsolidated rocks	Area with alluvial deposit	*Vineyards*
2(1)* and 2(2)*	Unconsolidated rocks	Alluvial fan	*Vineyards*
3	*Massive rocks*	Slope scarps	Irrelevant
4	Irrelevant	*Swampy areas*	Irrelevant
5	Unconsolidated rocks	Irrelevant	*Rivers*
6	Massive rocks	*Fluvial gorge*	Irrelevant
7	Irrelevant	Area with glacial deposits	*Apple orchards*
8	*Rigid-plastic rocks*	Slope scarps	Irrelevant

* Presence (1) / absence (2) of "DOC" areas.

the worse is the score given to the alternative (Patrono 1994; Patrono et al. 1995). Loss of value refers to the degradation due to the human activity, here established after adapting to a GIS an approach originally developed by Panizza et al. (1995) for the assessment of impacts on landforms. First the quality (Q) of each impacted landscape unit, representing a unique combination of the three indicator class values or landscape components, was assessed as follows:

$$Q = (L_g + L_o + L_p + L_e) \times C \tag{1}$$

where C = condition of preservation (0 < C ≤ 1); L_i = the level of interest of the following characterising factors: g = example for geomorphologic evolution (with specific focus also on the geologic and land-cover settings); o = object of educational purposes; p = paleogeomorphologic example; e = key role of geomorphology in an ecosystemic perspective. They all range between 0 and 1 (e.g., 0 means no interest, while 1 means a world-wide recognised situation).

The impact for each characteristic combination of indicator class values was expressed as a loss of natural quality as follows:

$$I_c = (\Sigma\ A_{ci} \times Q_{ci}) / (A_{totc} \times Q_{maxc}) \tag{2}$$

where c = considered combination of the three indicators; A_{ci} = impacted "c" areas; Q_{ci} = quality of the impacted "c" areas; A_{totc} = total "c" areas; Q_{maxc} = maximum quality encountered in "c" areas.

The impacts caused by the alternatives were evaluated in terms of space occupation effect (Patrono et al. 1995). The presence of infrastructures implies the occupation of identifiable areas, which were extrapolated directly from the project using spatial buffering techniques (Serra 1982). Space occupation areas were considered according to the characteristics of the project (e.g. see Lanzavecchia 1986). Once for each alternative the spatial distribution of the impacts was calculated, the impact maps were overlapped with the TMU map. In that manner it was possible to evaluate and quantify the presence / absence of impacts in each TMU. Once also a scientific value had been associated to each TMU, according to expression (1), the impact was evaluated using expression (2). Table 5 shows the data extracted to compute the impact on the criteria 2(1) and 5 (Tab. 6). The operation had to be repeated for all relevant combinations of indicators and, of course, for each of the five alternatives. Table 7 shows the results of impact calculations after normalisation (all impact values were divided by the maximum impact value). The values in the table represent the size of the impacts for each alternative. The impacts of the alternatives are expressed in quantitative form and they are considered as costs so that they are negative (i.e. the lower, in "absolute" terms, the better). As it can be seen in Table 7 comparing it with Table 6, some of the criteria originally selected are missing due to the absence of impact on that specific combination of indicator class values. The presence of tunnels had to be dealt with separately. Tunnels have no influence in terms of space occupation but excluding them would have meant to favour a priori the less "open" alternatives. Nevertheless tunnels produce a large amount of waste, which may not be important in terms of loss of resources, due to their unlimited abundance but it represents a very critical "dumping" problem. For this reason the impact was simply derived from the histogram of the map of each alternative by measuring the length of the tunnels and considering their characteristics constant for all the alternatives.

Table 7. Impact results (normalised).

Criterion	Alt. 1	Alt. 2	Alt. 3	Alt. 4	Alt. 5
4	-1.000	-0.419	-0.419	-0.419	-1.000
5	-0.954	-0.129	-1.000	-0.129	-0.787
1(1)	-0.854	-0.530	-0.842	-0.527	-1.000
1(2)	-0.012	-0.000	-0.012	-1.000	-0.000
2(1)	-1.000	-0.801	-1.000	-0.801	-0.824
Galleries	-0.451	-0.858	-0.447	-1.000	-0.451

6 APPLICATION OF MULTI-OBJECTIVE DECISION SUPPORT SYSTEMS (MODSS)

The number and variety of decision makers in EIA and the complexity of the information needed in environmental studies, call for forms of decision support which are more adequate and accessible than what is presently available. Multi-Objective Decision Support Systems can play an important role in environmental management. They represent ideal computer tools to emphasise the use of experimental results in complex decision environments involving many objectives and / or many disciplines (Janssen 1992).

In this study a multi-objective decision support system for environmental problems, the computer program DEFINITE (Janssen & van Herwijnen 1992), is used to assess the general impact produced by the five project alternatives and to reach a satisfactory ranking. According to the analytical methods built in DEFINITE, the decision process adopted in this study consisted of the following steps: (a) Input of an effect table representing the decision problem (Tab. 7); (b) Generation of a complete (/ incomplete) ranking of the alternatives; (c) Analysis of the sensitivity of the scores, expressing their uncertainty. The impact matrix shown in Table 7 was used as input to the multi-criteria analysis methods programmed in DEFINITE for ranking the alternatives. The procedures used to support the evaluations are described in the following paragraphs.

6.1 Weight assignment

The use of multi-criteria methods of analysis requires information on the relative importance of each criterion. Different weighting methods were used to transform qualitative priority statements into quantitative weights: (a) the Expected Value method, (b) the Random Value Method and (c) the Extreme Value Method (Janssen 1992). The importance weights so determined were used later in multi-criteria evaluation to rank the alternatives. Ranking methods treat the order of criteria importance established by the decision-maker as information on the unknown quantitative weights and try to make optimal use of that information. The criteria arrangement used in this analysis, from the most important (a) to the least important (d) (notice that there are some "as important as" situations - see also Table 6), was: (a) Criterion 1(1) and Criterion 2(1), (b) Criterion 4, (c) Criterion 5, (d) Criterion 1(2) and Tunnels.

The philosophy adopted in assigning a priority is clearly assuming an economic point of view in considering the presence of the alternatives in the landscape, where economic means a strong link between valuable cultivated land and the geomorphologic factor. On the other extreme there is the production of waste, represented by tunnels, and the loss of less valuable cultivated land.

6.2 Evaluation results

The problem addressed in the evaluation process is to judge the attractiveness of the alternatives. The methods used in this study were: (a) Weighted summation, (b) Electre, and (c) Expected Value (Janssen 1992). Table 7 was used as input to rank the alternatives and the results obtained are summarised in Table 8. Considering the extreme situations, alternative 2 represents definitely the best solution according to the assigned relative importance, while alternative 1 is constantly the worse. The other three alternatives seem quite similar and interchangeable, with a weaker dominance of alternative 4. It means that ranking modifications can be expected with small modifications of the importance values. Considerations about the probability of shifting positions due to small changes in the importance values are analysed in the sensitivity analysis, which follows.

Table 8. Result of the Multi-objective decision support system.

Multi-criteria method	Weighting procedure	Rank of Alternatives
Weighted Summation	Expected weights	2, 4, 5, 3, 1
Weighted Summation	Random weights	2, 4, 5, 3, 1
Weighted Summation	Extreme weights	2, 4, 3 & 5, 1
Electre	Expected weights	2, 3, 4 & 5, 1
Electre	Random weights	2, 3, 5, 4, 1
Electre	Extreme weights	2, 3, 4 & 5, 1
Expected Value	Expected weights	2, 4, 5, 3, 1
Expected Value	Random weights	2, 4, 5, 3, 1
Expected Value	Extreme weights	2, 4, 3 & 5, 1

6.3 Analysis of the sensitivity

Often, in EIA scores and priorities are uncertain and evaluation methods involve different assumptions. The sensitivity of the rankings obtained due to changes in evaluation methods, impact scores and weights can be analysed using DEFINITE. The sensitivity analyses performed were here as follows: (a) the least distance weight and (b) the weight intervals with reversal in the ranking of alternatives 3, 4 and 5. This procedure determines the intervals within which the rank order of a couple of alternatives is insensitive to changes in weights. This procedure in the (a) situation calculates the weights that are more similar to the assigned ones and in the (b) situation it calculates how a reversal of the rank order of two selected alternatives can be obtained. The results obtained can be summarised as follows: (a) comparing the results obtained by alternative 5 and alternative 4, and considering the relative importance originally assigned to the criteria, an improvement of weight on tunnels and non-DOC areas would definitely set alternative 5 as the second choice; (b) the same is obtained by alternative 3 when compared to alternative 4, when increasing the importance attributed to rivers and non-DOC areas; (c) alternative 3 and 5 are definitely very interchangeable; the only way to establish a leadership is to modify the roles of the first two criteria, originally considered "as important as". Whenever the impacts on the alluvial deposits (valley bottom) are considered less important than the impacts on the alluvial fans, alternative 5 represents a better choice. Viceversa alternative 3 becomes a better choice if a higher importance is given.

7 CONCLUSIONS

It has to be remarked the environmental impact assessment performed in this study is mainly focused on the impacts of the described project on the geomorphologic resources. Thus the conclusions reached are not taking into account the cost of the alternatives, the construction implications from the engineering point of view, the air pollutants effects on the roadside land-use / land-cover types, the cost of drilling tunnels and of disposing of the waste, etc. More considerations might change the final results and this represents the limitation of this study. EIA has to be always considered in an integrated perspective, which includes several disciplines to fully cover environmental and socio-economic implications of the intervention. This study is a specific contribution on the role of the geomorphologic component in EIA studies. The importance of geomorphology has been analysed in the suitability phase by studies (e.g. Panizza 1987; van Westen 1993; Cavallin et al. 1994), but more research is now needed for an operational evaluation phase.

The method described here seems to fill this gap still avoiding more sophisticated analyses and being a transparent and reliable tool, easy to implement in any GIS with an interface with Data-Base management system or with a spreadsheet. The role of subjectivity in establishing the values to be used in expression (1) and then in setting the relative importance for the criteria are part of the EIA game and cannot be avoided. The only way to limit its influence is to simplify and automates as much as possible the GIS analysis and that was one of the goals of this study.

Considering the results obtained, it is very important to notice in this application the presence of an optimal solution and of a constantly dominated alternative. The optimal solution is characterised by the best compromise length / presence of tunnels whose impact is the less important. From the engineering point of view it does not seem to be the best solution but this is something that was not to be evaluated here. The worse alternative is definitely the one that affects more riverbanks and vineyards. It is also the longest in terms of viaducts (bigger space occupation). The subjectivity involved in the MODSS application leave some uncertainty on the selection of the optimal alternative upon the remaining three solutions but with the aid of the sensitivity analysis it was possible to identify the reasons of this uncertainty and is now the task of the decision-maker to choose.

Concluding, it appears that the developed method satisfies well the need of representing the typology of the project and its impacts, affecting different geomorphologic settings, including the spatial relations project / environment. The loss of naturalness importance is clearly established and calculated. The integrated use of a GIS, of a relevant spatial databases and of a MODSS results critical.

ACKNOWLEDGEMENTS

Row data were provided by the Servizio Geologico, the Ufficio Valutazione d'Impatto Ambientale and the Ufficio Trasparenza of the Provincia Autonoma di Trento ("Convenzione n. racc. 020311, Provincia Autonoma di Trento-ITC"). This research was also funded by the EC HCM project (contract ERBCHRXCT 930311).

REFERENCES

Bartolomei, G., Corsi, M., Dal Cin, R., D'Amico, C., Gatto, G. O., Gatto, P., Nardin, M., Rossi, D., Sacerdoti, M. & Semenza E. 1969. *Note illustrative della carta geologica d'Italia alla scala 1:100,000, Foglio 21, Trento.* Ercolano (Napoli): Poligrafica & Cartevalori.
Beinat, E. 1995. Analysis and decision support techniques for environmental impact assessment (EIA). In M. Marchetti, M. Panizza, M. Soldati & D. Barani (eds), Geomorphology and Environmental Impact Assessment. Proceedings of the 1st and 2nd workshops of a "Human Capital and Mobility" project. *Quaderni di Geodinamica Alpina e Quaternaria* 3: 43-62.
Carver, S.J. 1991. Integrating multi-criteria evaluation with geographical information systems. *Int. Journ. GIS* 5(3): 321-339.
Cavallin, A., Marchetti, M., Panizza, M. & Soldati M. 1994. The role of geomorphology in environmental impact assessment. *Geomorphology* 9: 143-153.
Comprensorio Valle dell'Adige - Carta geomorfologica per la pianificazione territoriale ed urbanistica 1:10,000. Provincia Autonoma di Trento (unpubl.).
Fabbri A.G., Patrono, A. & Veldkamp J.C. 1995. A Case Study in Environmental Impact Assessment. In M. Marchetti, M. Panizza, M. Soldati & D. Barani (eds), Geomorphology and Environmental Impact Assessment. Proceedings of the 1st and 2nd workshops of a "Human Capital and Mobility" project. *Quaderni di Geodinamica Alpina e Quaternaria* 3: 109-122.
ITC 1996. *ILWIS 2.0 Reference Guide.* Enschede: International Institute for Aerospace Survey and Earth sciences (ITC).
Janssen, R. 1992. *Multiobjective Decision Support for Environmental Management.* Dordrecht: Kluwer Academic Publ.
Janssen, R. & van Herwijnen M. 1992. *DEFINITE. Decisions on a FINITE set of alternatives.* Dordrecht: Kluwer Academic Publ.
Landolt, E. 1977. Ökologische Zeigerwerte zur Schweizer Flora. *Ber. Geobot. Inst. ETH* 64: 64-207.
Lanzavecchia, S. 1986. *Guida pratica alle Valutazioni di Impatto Ambientale - Corso di formazione sulla valutazione ambientale.* Roma: Quaderni CNR, Settore Energia.
Marchetti, M., Panizza, M., Soldati M. & Barani D. (eds) 1995. Geomorphology and Environmental Impact Assessment. Proceedings of the 1st and 2nd workshops of a "Human Capital and Mobility" project. *Quaderni di Geodinamica Alpina e Quaternaria* 3: 1-197.
Meijerink, A.M.J. 1988. Data acquisition and data capture through terrain mapping units. *ITC Journ.* 1: 23-44.
Nijkamp, P. & H.J. Scholten 1993. Spatial information systems: design, modelling, and use in planning. *Int. Journ. GIS.* 7(1): 85 - 96.
Panizza, M. 1987. Geomorphological hazard assessment and the analysis of geomorphological risk. In V. Gardiner (ed), *International Geomorphology*: Part I: 225-229. New York: Wiley & Sons.
Panizza, M., Marchetti, M. & Patrono A. 1995. A proposal for a simplified method for assessing impact on landforms. *ITC Journ.* 4: 324.
Patrono, A. 1994. A Study in Environmental Impact Assessment (EIA). Integrated methodology to assess and predict the ecological impact of a motorway project in the province of Trieste, Italy. Enschede: ITC - MSc thesis (unpubl.).
Patrono, A. 1996. Synergism of remotely sensed data for land-cover mapping in heterogeneous alpine areas. An example combining accuracy and resolution. *ITC Journ.* 2: 101-109.
Patrono, A., Fabbri, A.G. & Veldkamp J.C. 1995. GIS analysis in geomorphology for environmental impact assessment studies. *ITC Journ.* 4: 347-353.
Peer, T. 1990. *I Biotopi in Alto Adige.* Bolzano: Athesia.

Pereira, J.M.C. & Duckstein L. 1993. A multiple criteria decision-making approach to GIS-based land suitability evaluation. *Int. Journ. GIS* 7(5): 407-424.

Prosser, G. & Selli L. 1991. Thrusts of the Mezzocorona-Mendola pass area (Southern Alps, Italy): structural analysis and kinematic reconstruction. *Boll. Soc. Geol. It.* 110: 805-821.

Provincia Autonoma di Trento. Carta geologica 1:25,000. Trento: Servizio Geologico Provincia Autonoma di Trento (unpubl.).

Rivas, M.V., Rix, K., Frances, E., Cendrero, A. & Brunsden D. 1995. The use of indicators for the assessments of environmental impacts on geomorphological features. In M. Marchetti, M. Panizza, M. Soldati & D. Barani (eds), Geomorphology and Environmental Impact Assessment. Proceedings of the 1st and 2nd workshops of a "Human Capital and Mobility" project. *Quaderni di Geodinamica Alpina e Quaternaria* 3: 157-180.

Scholten, H.J. & Stillwell J.C.H. (eds) 1990. *Geographical Information Systems for Urban and Regional Planning*. Dordrecht: Kluwer Academic Publ.

Serra, J. 1982. *Image Analysis and Mathematical Morphology*. London: Academic Press.

Westen, van C.J. 1993. *GISSIZ. Training Package for Geographic Information Systems in Slope Instability Zonation*. Enschede: ITC.

Color plates Chung & Fabbri

Figure 1. Earth and debris flows, or EDF landslides, which occurred prior to 1955 shown in black and EDF landslides which occurred between 1956-1993 shown in red in the Fabriano study area, central Italy. The black pre-1955 occurrences have been used to obtain the membership functions described in (5.5) and subsequently used to obtain all predictions using fuzzy operators, including Figures 3 and 4. They are also used to obtain the success rates in Table 1a and the corresponding Figure 5a. EDF landslides which occurred during 1956-1993 in red have been used to perform the cross-validation test, and to obtain the prediction rates in Table 1b and the corresponding Figure 5b.

Figure 2. Based on pre-1955 EDF landslide occurrences shown in black in Figure 1 and the six layers of 1955 spatial data described in Table 2, predicted EDF landslide hazard map in Fabriano study area, central Italy, was generated by using Equation (4.8) of the direct procedure described in § 4.3. Colours represent area percentages occupied by the sorted predicted values: 1) 0-5%; 2) 5-10%; 3) 10-15%, high; 4) >15%.

Figure 3. Based on pre-1955 EDF landslide occurrences (black in Fig. 1) and the six layers of 1955 spatial data (Tab. 2), the predicted EDF landslide hazard map in Fabriano study area, central Italy, was generated by using Equation (6.1) of the fuzzy minimum operator described in Section 6.1. Colours represent area percentages occupied by the sorted predicted values: 1 (0-5%); 2 (5-10%); 3 (10-15%, high); 4 (15-20%); 5 (20-25%, medium); 6 (25-30%); 7 (30-35%, low); 8 (35-40%); 9 (40-45%); 10 (45-50%); 11 (>50%).

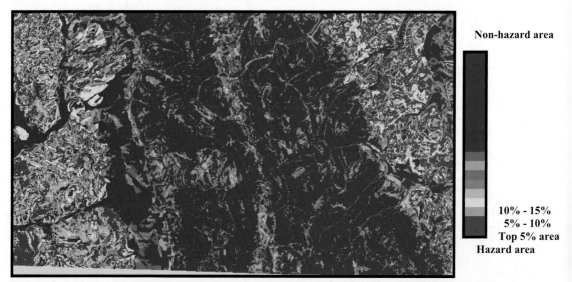

Figure 4. Based on pre-1955 EDF landslide occurrences (black in Fig. 1) and the six layers of 1955 spatial data (Tab. 2), the predicted landslide hazard map in Fabriano study area, central Italy, was generated by using Equation (6.4) of the fuzzy algebraic sum operator (Section 6.4). Colours represent represent area percentages occupied by the sorted predicted values: 1 (0-5%); 2 (5-10%); 3 (10-15%, high); 4 (15-20%); 5 (20-25%, medium); 6 (25-30%); 7 (30-35%, low); 8 (35-40%); 9 (40-45%); 10 (45-50%); 11 (>50%).

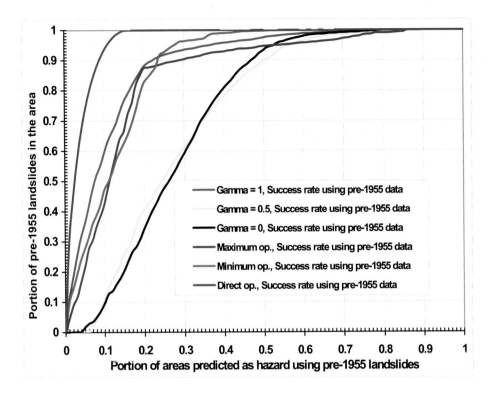

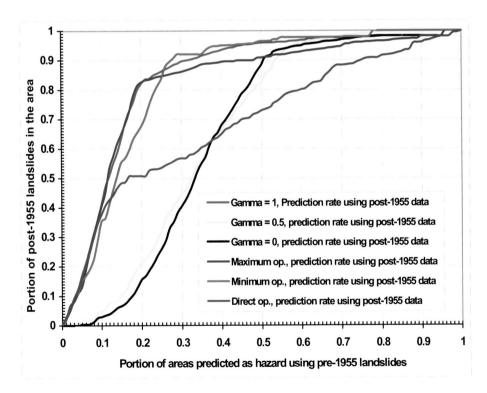

Figure 5. (a) Success rates for pre-1955 EDF landslide occurrences (training data) based on the membership functions and the fuzzy set models using pre-1955 data, Fabriano area, central Italy. The corresponding tabular data are shown in Table 1a. (b) Prediction rates for 1956-1993 EDF landslide occurrences based on the membership functions and the fuzzy set models using pre-1955 data, Fabriano area, central Italy. The corresponding tabular data are shown in Table 1b.

Color plates Patrono et al.

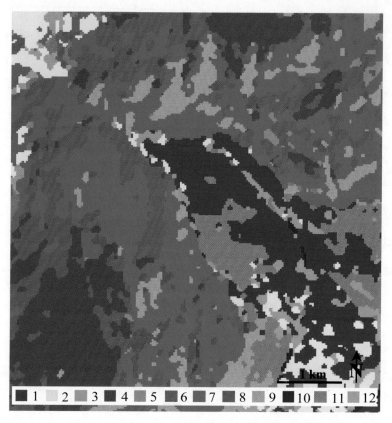

Figure 4. Land-use / land-cover map (for the legend specifications see Table 4).

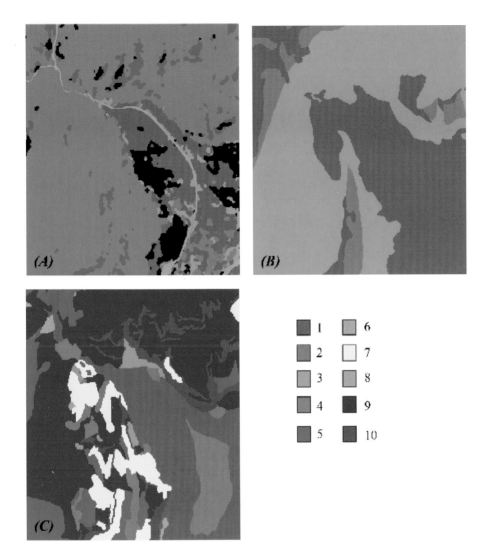

Figure 5. Maps of the three reclassified indicators: (a) geomorphology, (b) geology, (c) land-use / land-cover. The values of the legend refer to: (a) "R.V." in Table 4, (b) "R.V." in Table 2, (c) "CL.V." in Table 3.